Paper Panzers the Unfinished Projects

A rigid system for development and procurement of weapons systems had been established for the German army long before the Nazi era started in 1933. The troops were to send in requests for advanced weapon designs to their branch inspectorate. Controlled by budgetary constraints, the inspectorate (In 6 for tanks) eliminated crackpot ideas and Xmas wish lists. For worthwhile ideas, In 6 created performance specifications and sent them to the branch ordnance department (Wa Pruef 6 for automotive). Wa Pruef 6 then created a design specification and selected two or three private companies to develop conceptual designs. 1/10-scale drawings and wooden models were created by the companies (such as Krupp, Rheinmetall, Daimler-Benz, and M.A.N.) for Wa Pruef 6. These conceptual designs were then shown to In 6 to determine if they met their needs and to approve the next steps for procuring test Panzers. The system was very well controlled, because In 6 was not allowed to design and Wa Pruef 6 had no money. Therefore, Panzers designed up to 1939 were created as basic workhorses for employment in a wide range of tactical functions.

This well-controlled system broke down at the start of the war, when Wa Pruef 6 themselves started creating new designs. Being engineers, they went for the newest high technology engines, semi-automatic transmissions, complex steering units, and torsion bar suspensions - none of which had been proven reliable or were needed to fulfil basic tactical needs. Then, companies like Porsche and Daimler-Benz started their own conceptual designs without guidance from In 6 or Wa Pruef 6.

Having encountered British, French, and Russian tanks with thick armor at the start of the war, Germany embarked on an uncontrolled arms race for bigger guns and thicker armor. Specifications based on sound tactical principles were discarded, and politicians became involved in technical decisions.

Totally out of control, the engineers' design efforts were wasted on numerous conceptual design projects. Quickly thrown together, these projects were just as rapidly abandoned when proven to be incompatible with the reality of their industrial production capabilities.

However, these misguided efforts have left us with a very interesting array of designs. They demonstrate what can be created by engineers pushing on the state-of-the-art high technology envelope. Many of these projects, like the Pz.Kpfw.Loewe (Lion) and the Baer (Bear) never got off the drawing board. Others, like the Jagdpanzer 38 D, almost made it into series production at the end of the War.

There are hundreds of surviving original conceptual design drawings - way too many to cover in a single Panzer Tracts volume. Therefore, this first Paper Panzers edition contains only the Pz.Kpfw., Sturmgeschuetz, and Jagdpanzer projects. Future editions will cover the reconnaissance vehicles, the Flakpanzers, and the self-propelled anti-tank guns and artillery.

Pz.Kpfw.IX and X were not actual conceptual design projects. These sketches were published as misinformation in Heft 3 1944 of "Signal" to fool the Allies on plans for the next generation of super Panzers.

→ „Nähere Angaben fehlen noch." Beinahe wie Schiffe zu Lande sehen diese neuen Entwurfe der Panzer aus. Rundungen und Schragungen sollen auftreffende Granaten abgleiten lassen

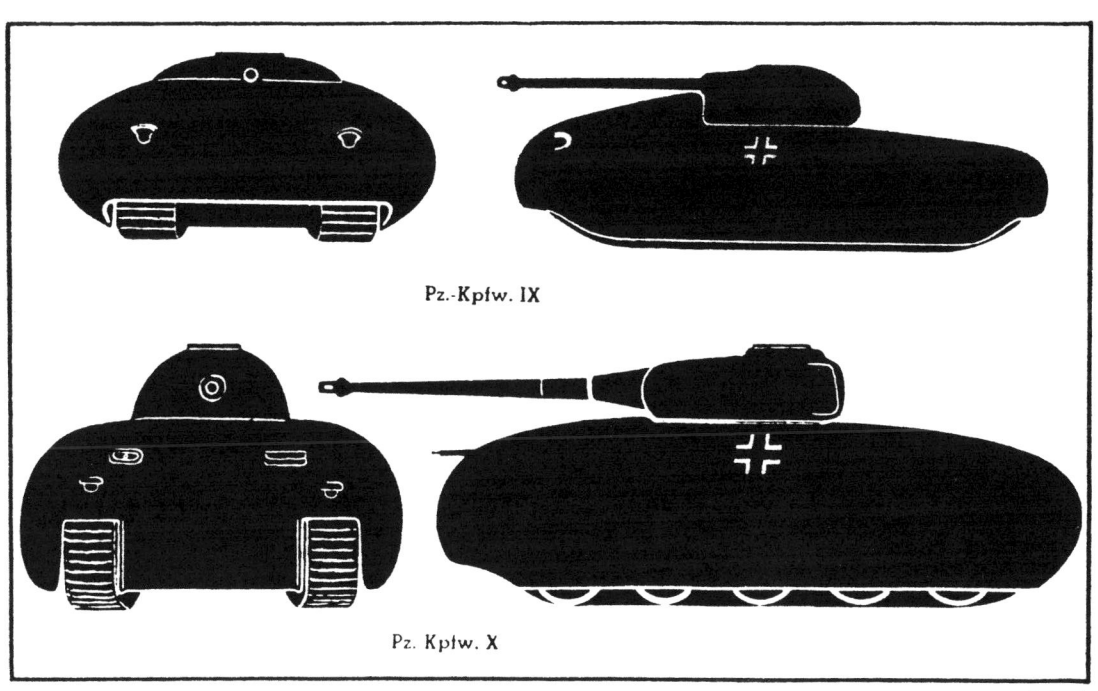

Pz.-Kpfw. IX

Pz. Kpfw. X

VK 20 SERIES

The search for a replacement for both the Pz.Kpfw. III and IV began in 1938. This new series of tanks was designated as VK 20.01. The VK stood for Vollketten (fully tracked), the 20 for the weight class in tons, and the 01 for the first model in the series. The first contract for a detailed design was given to Daimler-Benz. In October 1939, Daimler-Benz started to design their own tank after receiving permission to prepare designs without outside interference. At the beginning of the War in September 1939, Krupp started to design an improved Pz.Kpfw.IV. This project was canceled in May 1940 and Krupp began detailed designs on a new tank in the 20 ton class. Engineers at both Daimler-Benz and Krupp favored leaf spring suspensions and were opposed to torsion bars. Therefore, Maschinenfabrik Augsburg-Nuernberg (M.A.N.) was pulled into the competition to design a new tank in the 20 ton class with a torsion bar suspension.

VK 20.01 FROM DAIMLER-BENZ

Daimler-Benz was awarded a contract for creating the detailed design for a new tank chassis in the 20 ton class, known as the VK 20.01 (III). Unlike the Z.W.40 project which was based on modifying a normal Pz.Kpfw. III chassis, the VK 20.01 (III) was a complete departure from previous designs. By 14 December 1938, Daimler-Benz had completed a design project with the 6 cylinder Maybach HL 116 motor developing 300 metric horsepower at 3300 rpm. The VK 20.01 (III) was one of the first tanks designed with a Schachtellaufwerk (interleaved roadwheel suspension) with torsion bars.

In October 1939, Daimler-Benz received permission from the Generalbevollmaechtigen (commission for standardization of automotive designs) to independently design a tank without interference from Wa Pruef 6. The new tank was initially designated by Daimler-Benz as the GBK (Kampfwagen des Generalbevollmaechtigen) and only later as the VK 20.01 (D).

On 15 November 1939, Daimler-Benz decided on the following design features for the GBK. A Wilson transmission was to be designed by CKD. As a backup, a manually shifted transmission similar to the Zahnradfabrik SSG 77, but adapted to the higher torque of the Daimler-Benz MB 809 diesel engine, was to be developed by Daimler-Benz. A leaf spring design was to be installed in the new GBK. Roadwheels of 680 mm diameter (drawing No. 021 B 10814) were initially considered for the GBK project.

The design for the 12 cylinder MB 809 diesel motor, rated at 350 metric horsepower was completed in June 1940. The first motor ran on a test stand in February 1941, followed by its acceptance test on 12 March 1941. On 21 March 1941, this motor arrived in Berlin-Marienfelde to be installed in the experimental chassis. Driving tests were conducted at the assembly plant and in Kummersdorf.

On 1 August 1941, the VK 20.01 (D) was reported to weigh 22.25 metric tons with a maximum speed of 50 km/hr. It had a leaf spring suspension supporting the 700 mm diameter roadwheels, which ran on 440 mm wide Kgs 62/440/120 track.

As reported on 22 December 1941: Based on experience in the Russian campaign, the new tank just developed by Daimler-Benz was now obsolete. Utilizing the already developed tank, studies were being conducted to develop a new design (the Panther) with thicker armor and heavier armament.

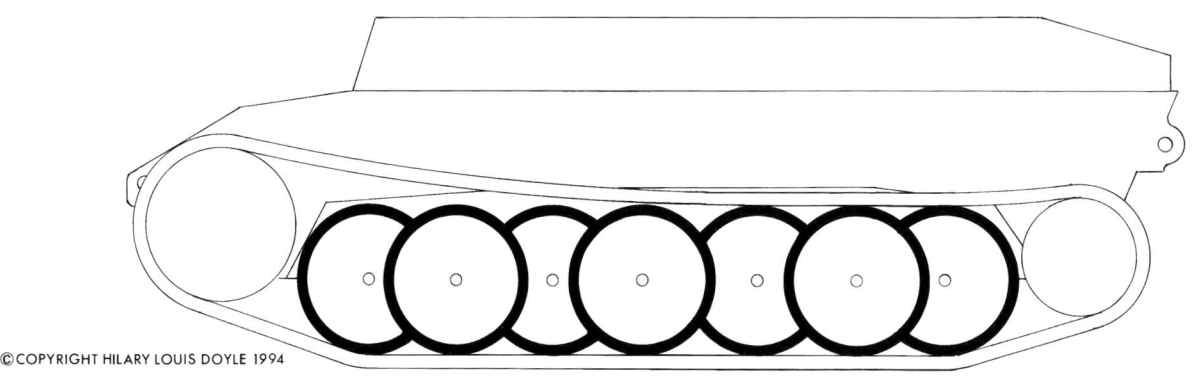

The Daimler-Benz comceptual design for their VK 20.01 (D)

VK 20 SERIES FROM KRUPP

At a meeting on 15 September 1939, Wa Pruef 6 and Krupp discussed the design of a new chassis, the VK 20.01 (IV), as a further development of the Pz.Kpfw.IV series. The engine compartment and the Maybach HL 116 engine were to be adopted from the VK 20.01 (III). Maximum speed was to be 42 kilometers per hour, the same as the Pz.Kpfw.IV Ausf.C. The suspension was selected to support a vehicle weight of 20 metric tons with a hull width of 1820 mm and an overall width of 2900 to 2950 mm. The Schachtellaufwerk (interleaved roadwheel suspension) from the VK 20.01 (III) couldn't be used, since it resulted in an overall width of about 3040 mm and exceeded the weight specification. Krupp proposed a six- wheel suspension with leaf springs which could use components from the normal eight-roadwheel suspension for the Pz.Kpfw.IV.

At a meeting on 28 October 1939 with Wa Pruef 6, a diameter of 630 mm was set for a six-roadwheel suspension for the VK 20.01 (IV). New Kgs 61/400/120 tracks were to be utilized. The designation for VK 20.01 (IV) evolved to VK 20.01 (BW) in November and then was changed to B.W.40 in December 1939.

On 13 December 1939, the frontal armor for the B.W.40 was increased from 30 mm to 50 mm thick. The basic hull shape for the B.W.40 was the same as the Pz.Kpfw.IV Ausf.D. On 4 January 1940, Wa Pruef 6 notified Krupp that two armor hulls and one soft steel hull were to be produced for the three B.W.40 experimental chassis. Three superstructures and rear decks were to be made from soft steel.

On 16 May 1940, Wa Pruef 6 informed Krupp that in consideration of the wartime situation, the B.W.40 project was to be shelved.

Also in May 1940, Krupp initiated a new design in coordination with Wa Pruef 6, the VK 20.01 (K) with either the Maybach HL 116 or HL 115 motors. As recorded in Krupp's 1939/1940 fiscal year report, preliminary proposals had been completed for the VK 20.01 (K) with 5 cm armament and heavier armor. A full-scale wooden model was being completed, and a detailed design had been started. The experimental turret for the VK 20.01 was being designed by Krupp with 5 cm Kw.K. L/42 armament. Frontal armor was to be 50 mm thick, side and rear armor 30 mm thick.

By 24 October 1940, Krupp had received contracts for three VK 20.01 (K) developmental chassis and by 12 November 1940 a contract for production of a 0-Serie of 12 VK 20.01 (K) complete with 5 cm gun turrets. Wa Pruef 6 had awarded contracts to M.A.N. and Daimler-Benz for the design of new tanks with 7.5 cm gun turrets in the 20 ton class.

In March 1941, Krupp proposed to complete three developmental VK 23.01 (K) chassis and six experimental VK 23.01 (K) with torsion bars. The VK 23.01 (K) was to have the power train components designed by Kniepkamp and developed by M.A.N.

On 18 April 1941, Wa Pruef 6 informed Krupp that they were to concentrate on development of a new submersible turret with a 5 cm Kw.K. L/60 gun for the VK 20.01 (K). The experimental turret for the new VK 20.01 (K) was to be completed by 1 February 1942 directly followed by production of 12 turrets for the 0-Serie.

By July 1941, an additional contract had been awarded to Krupp for three developmental VK 23.01 (K) chassis in soft steel without turrets. The previous contract for a 0-Serie of 12 was revised to specify two redesigned variants: six VK 20.02 (K) and six VK 23.01 (K). The six armored VK 20.02 (K) were to have 5 cm Kw.K. L/60 guns in their turrets.

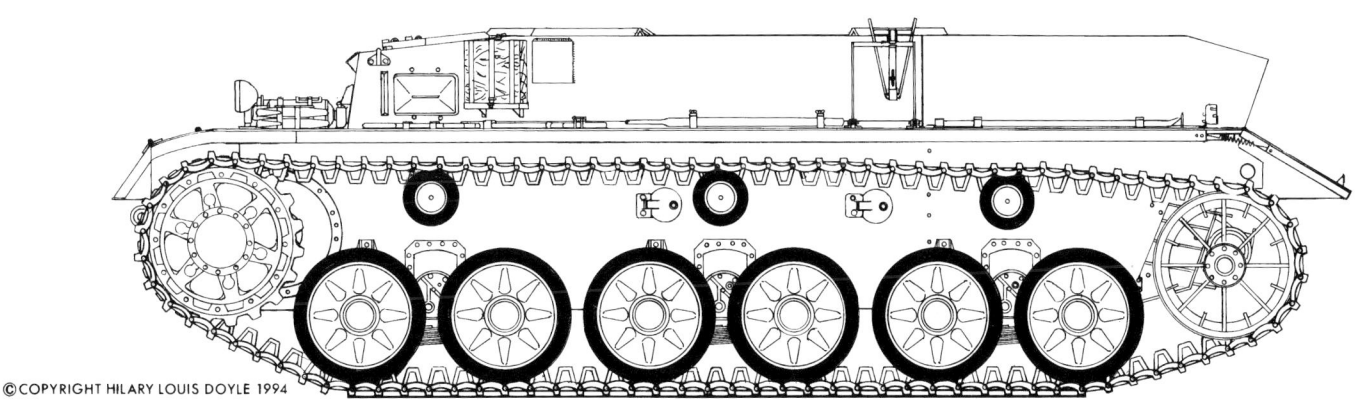

The conceptual design for the B.W.40 from Krupp

The conceptual design for the VK 20.01 (K) from Krupp

Both the VK 20.01 (K) and VK 20.02 (K), designed for a maximum speed of 56 km/hr, had the same leaf spring suspension with six roadwheels (700 mm diameter) running on 450 mm wide Kgs 62/450/120 track. The total weight of a complete VK 20.01 (K) including a turret was calculated to be 21.5 metric tons, compared to the VK 20.02 (K) at 23 metric tons.

To meet the requirement for standardization demanded by Kniepkamp, the VK 23.01 (K) was to have a Schachtellaufwerk designed by M.A.N. with six 880 mm diameter roadwheels, a torsion bar suspension, and 474 mm wide Kgs 63/474/110 tracks. The first VK 23.01 (K) developmental chassis with torsion bar suspension, outfitted for total submersion, etc., could be delivered, at the earliest, about 1 October 1942.

In designing their VK 23.01 (K), Krupp contacted Zahnradfabrik Friedrichshafen and Maybach on 19 September 1941 in order to obtain the latest installation drawings of transmissions for the Dreiradien-Lenkgetriebe (triple radius steering gear) designed by M.A.N.

In a report written in January 1942, Woelfert related how the design and completion of the VK 20.02 (K) had frequently been delayed due to the following reasons: Initially a 5 cm turret was specified with an inner turret ring diameter of 1350 mm without a traversing floor. Then, on 10 October 1941, a 5 cm turret with an inner turret ring diameter of 1400 mm was required, to allow the option of mounting a 7.5 cm gun turret. Finally, the 7.5 cm Einheitsturm (standard turret) with 7.5 cm Kw.K.44 was specified, at first with an inner turret ring diameter of 1560 mm, later increased to 1600. As a result of continuously increasing the turret ring diameter, the hull width was expanded from 1600 to 1650 mm and the hull (originally 400 mm shorter) had to be repeatedly lengthened. Also, the ammunition storage was fundamentally changed. Along with this came the requirement to slope the superstructure walls, especially the driver's front plate.

In a meeting on 17 December 1941 with Krupp, Oberst Fichtner (head of Wa Pruef 6) expressed his position on tank development as follows: Against the advice of Wa Pruef 6, higher authority had decided that the weight class for the future tank should be 30 tons and not the 24 ton tank proposed by Wa Pruef 6. Fichtner was opposed to this decision. In his opinion, time would be lost since the 30 ton tank had yet to be developed, whereas development of the 24 ton tank was almost completed. Also, the heavier 30 ton tank would result in a lower number produced. Not all assembly firms were in a position to manufacture a 30 ton tank, and larger facilities were also needed. In regard to tactical employment, engineer bridges couldn't support the weight of a 30 ton tank. Therefore, a 30 ton tank depended on submerged river crossing, but it was known that this design problem hadn't been satisfactorily solved. Wa Pruef 6 had been ignored and Reichsminister Todt had declared that a 30 ton tank must quickly be developed and produced.

At the end of December 1941, the contract for design and fabrication of the six VK 20.02 (K) was halted to allow accelerated completion of other tasks. Also in December 1941, the contracts for the VK 23.01 (K) with torsion bars were canceled so that other urgent design work would be quickly completed.

VK 20 SERIES FROM M.A.N.

Having been frustrated in the attempts to get Daimler-Benz and Krupp to design tanks with torsion

The conceptual design for the VK 20.02 (M) from M.A.N.

bar suspensions, Wa Pruef 6 turned to M.A.N. early in 1940. Layout drawings of the VK 20.01 (M) with a Schachtellaufwerk cushioned by torsion bars had been completed by M.A.N. by 10 October 1940. Due to the interference caused by mounts for the steering gear and the engine, torsion bars were replaced by coil springs on the first and last stations.

M.A.N. was awarded a contract by Wa Pruef 6 to design an improved version, the VK 20.02 (M). By February 1941, M.A.N. had completed preliminary drawings for fitting the SMG 91 transmission designed by Zahnradfabrik Friedrichshafen to their triple radius steering gear.

On 18 August 1941, M.A.N. reported that the VK 20.01 (M) chassis was already being assembled but a transmission was still needed for it to be completed. The three developmental VK 20.02 (M) chassis and 12 experimental VK 20.02 (M) chassis that had been contracted by Wa Pruef 6 were still being designed. The VK 20.02 (M) was to have a Schachtellaufwerk with six 880 mm diameter roadwheels, a torsion bar suspension, and 474 mm wide Kgs 63/474/110 tracks. The automotive drive train consisted of a Maybach HL 90 motor, a Maybach OG 32 6 16 or Zahnradfabrik SMG 91 transmission, and a M.A.N. Dreiradien-Lenkgetriebe (triple radius steering gear). Armor protection for the hull consisted of 50 mm front plates, 40 mm side and rear plates, and 14.5 mm plates for the roof and belly.

In response to reports from the Eastern Front on the success of sloped armor on Russian tanks, M.A.N. had redesigned the hull for the VK 20.02 (M) with sloping armor (drawing No. Tu 13947 dated 25 November 1941). The hull front, still 50 mm thick, was set at an angle of 55 degrees. The 40 mm thick upper hull side was set at 40 degrees, the 40 mm lower hull side remained at 0 degrees, but the 40 mm hull rear was canted at an angle of 30 degrees.

A VK 24.01 (M) was mentioned by M.A.N. in their postwar response to questions on what had influenced the design of the Panther, but no details have been found.

The VK 20.02 (M) redesigned with sloped armor (based on drawing Tu 13947 dated 25Nov41)

Panzerkampfwagen III Ausf.K (8./Z.W.)

At a meeting on 15 December 1941, the Heereswaffenamt informed Krupp that they were to prepare drawings to mount the B.W. (Pz.Kpfw.IV) turret on the 8./Z.W. (Pz.Kpfw.III). Preliminary drawings were completed by 27 December, and Krupp sent six sets of drawings for outfitting the Pz.Kpfw.III with the 7./B.W. turret to the armor supplier on 15 January 1942. Based on the difficulties that would be encountered in modifying the Panzer III to mount the Panzer IV turret (large increase in weight, new suspension, new track, changed center of gravity, etc.), Hitler decided against this change and to continue producing the Panzer III with the 5 cm Kw.K. On 19 March 1942, Krupp recorded that OKH had decided not to produce the Ausf.K with the 7.5 cm Kw.K. 40.

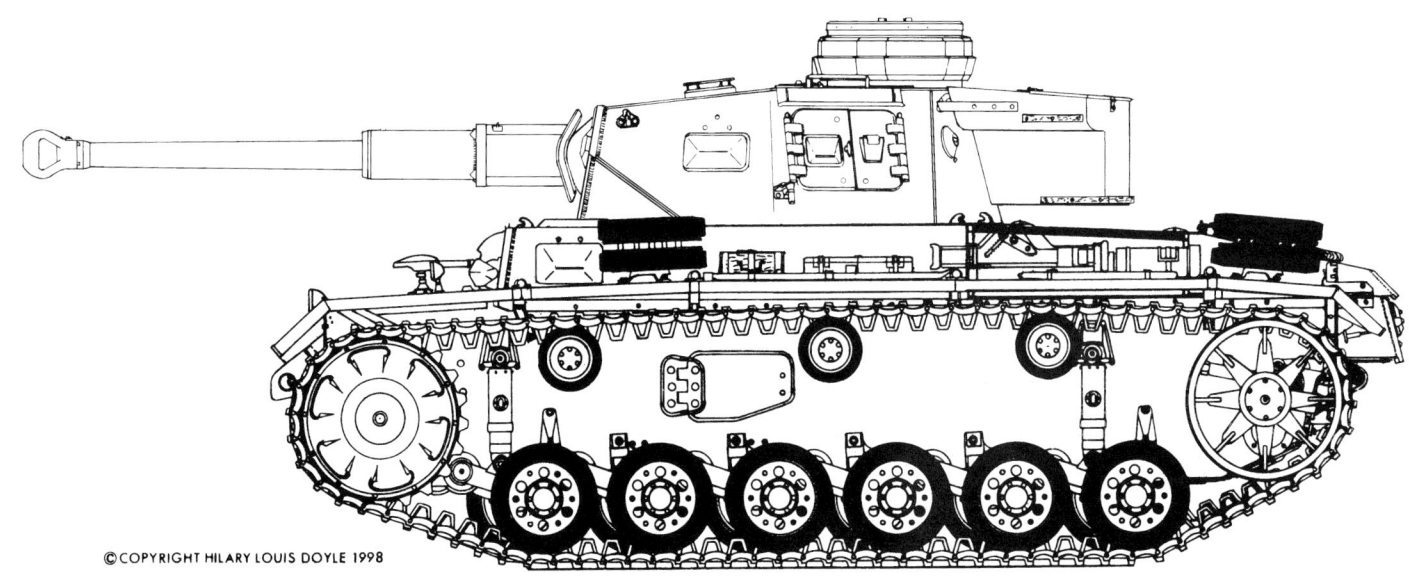

4./Z.W. (Pz.Kpfw.III Ausf.E) Fahrgestell fuer 3./B.W. (Pz.Kpfw.IV Ausf.C)

The Pz.Kpfw.III Ausf.K was not the first attempt to mount a turret with a 7.5 cm gun on a Pz.Kpfw.III chassis. In regard to the further development of the B.W. chassis for the Pz.Kpfw.IV series, Wa Prw 6 informed Krupp on 1 June 1937: "The necessity to standardize the Pz.Kpfw.chassis results in the discontinuance of the B.W. chassis at the end of the 2./B.W. (Pz.Kpfw.IV Ausf.B). Plans are to use the currently under development 4./Z.W. chassis with altered hull for the 3./B.W."

On 2 May 1938, Herr Woelfert (head engineer for Panzer automotive design at Krupp) stated that new production contracts (but no design contracts) were being awarded to Krupp. "Originally the Daimler-Benz-Einheitsfahrgestell (standard chassis) was to follow the 2./B.W. Serie. Setbacks have resulted from using newly designed components that weren't sufficiently tested for Panzers. Among these are the three-stage steering unit from Zahnradfabrik Friedrichshafen, the semi-automatic transmission from Maybach, and rubber-cushioned tracks. Experience with these gear boxes and tracks on Zugkraftwagen (three-quarter tracked towing vehicles) is insufficient for direct adaptation to Panzers. Since the first 4./Z.W. trial chassis is just now being sent to Kummersdorf for testing, this chassis won't enter series production in the foreseeable future. Large contracts are being distributed for the 3., 4., and 5.Serie Pz.Kpfw.IV with the B.W. chassis instead of the Daimler-Benz Einheitsfahrgestell."

Original 9./B.W. (Panzerkampfwagen IV Ausf.H)

On 31 December 1942, Wa Pruef 6 informed Krupp that the Generalstab had decided to modernize the B.W. hull with a sloped glacis as proposed by Krupp in drawing W1462. The weight would increase by 880 kg with the 50 mm sloped glacis, 80 mm hull front, and 20 mm forward belly plate. At the same time, wider tracks and three instead of two tandem roadwheels were to be mounted. Later, a suspension with Russenrollen (rubber-cushioned steel-tired roadwheels) would be considered.

On 5 February 1943, Krupp informed Wa Pruef 6 that a requirement to increase the sides and turret rear to 45 mm (in response to Russian anti-tank rifles) would increase the weight to 27.2 metric tons. Mounting the intended 560 mm wide Winterketten to lower the ground pressure to 0.72 kg/cm² would increase the overall weight to 28.2 tons. This would probably result in the failure of rubber-tired roadwheels and also cause steering and braking problems.

The future of the Pz.Kpfw.IV series was decided in a meeting of the Panzerkommission on 17 February 1943. The Adolf-Hitler Programm called for the Pz.Kpfw.IV production rate to more than double by October 1943, with an additional 1200 completed by the Spring of 1944. In order to meet these production goals, the assembly and steel firms insisted that the Pz.Kpfw.IV design needed to remain unchanged. The disadvantages of the new design outweighed the advantages, and the Panzerkommission decided to recommend to Hitler that the new Ausf. with sloped armor and strengthened belly and roof be dropped.

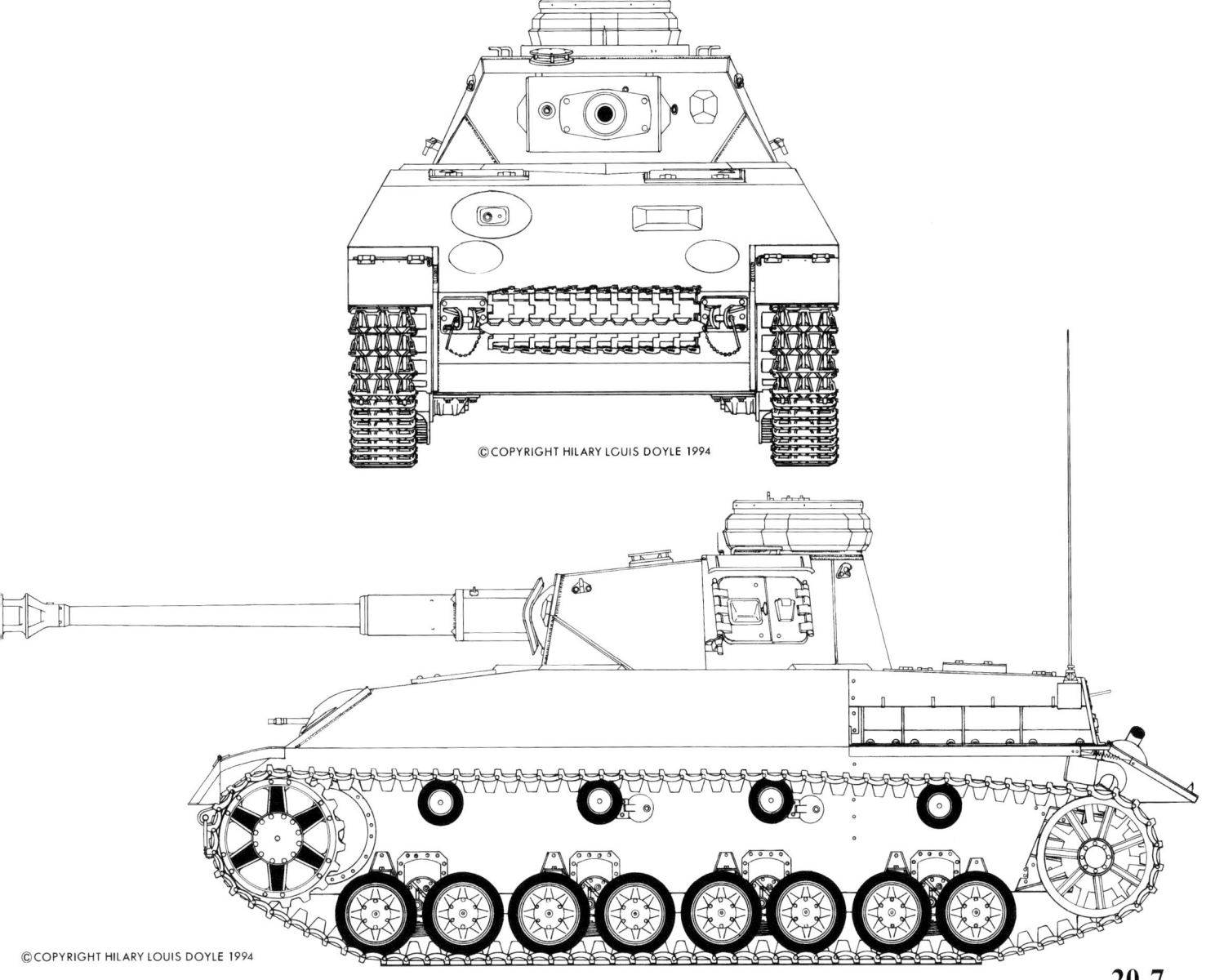

Panzerkampfwagen auf Einheitsfahrgestell III/IV
Ausf. A

Instigated by the requirement from Hauptdienstleiter Saur of the Speer Ministry that medium and heavy Panzers be standardized, the Panzerkommission met on 4 January 1944 to decide on the specifications for a combination Pz.Kpfw. III/IV. This same design was also to be used as the chassis for a leichte Panzer Jaeger (later renamed Panzer IV lang (E). The drive train was to retain the Maybach HL 120 TRM engine and cooling system from the Pz.Kpfw.IV in combination with the SSG 77 transmission, steering gear, and reinforced final drives and sprocket from the Pz.Kpfw.III. The suspension was to consist of six 660 mm diameter rubber-saving steel-tired roadwheels, mounted in pairs with leaf springs. New symmetrical 540 mm wide track was selected with a center guide, consisting of paired cast and formed links (similar to Tiger II track). The idler wheel was retained from the Pz.Kpfw.IV, but the adjusting mechanism had a wider span and was strengthened and simplified. The driving range was significantly increased by adding a 300 liter fuel tank in the engine compartment.

Front hull armor was redesigned to come to a point with sloped 60 mm thick glacis plate at 60 degrees and 60 mm lower hull plate at 45 degrees. The 80 mm thick driver's front plate was also well sloped at 50 degrees, and the 30 mm thick side of the superstructure sloped at 36 degrees. Schuerzen (steel armor skirts) were to be made out of 5 mm thick, flat interchangeable pieces.

Turret design remained practically the same as the Pz.Kpfw.IV Ausf.J. However, the slip-ring electrical contacts were to be dropped. The flexible electric cable, providing power to the turret, restricted the traverse to 270 degrees to either side. Ammunition stowage was increased to 100 rounds for the 7.5 cm Kw.K.40 L/48 along with 3150 rounds for the machinegun.

In March 1944, Wa Pruef 6 awarded contracts for the production of three trial Pz.Kpfw.IV auf Einheitsfahrgestell (standard chassis). A decision was made in June 1944 to start Pz.Kpfw.III/IV series production at Krupp-Grusonwerk in February 1945. On 12 July 1944, Krupp was informed that the Pz.Kpfw.III/IV was dropped and only the Panzerjaeger mit L/70 produced (Refer to Panzer IV/70 E).

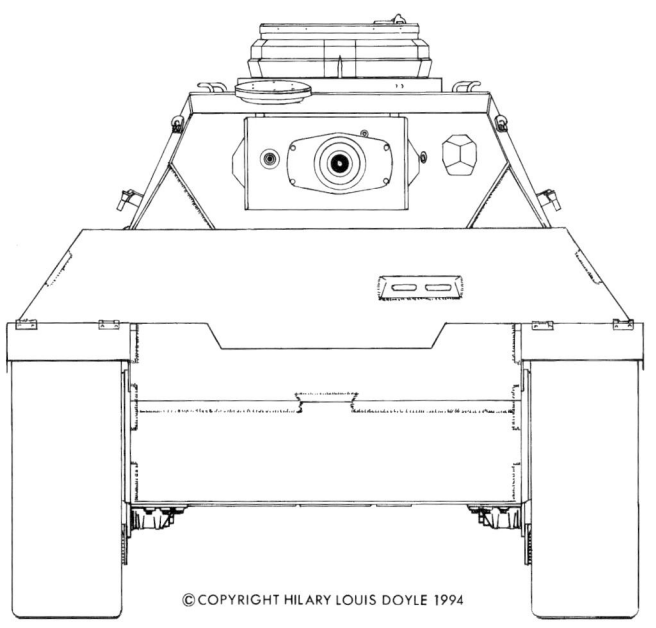
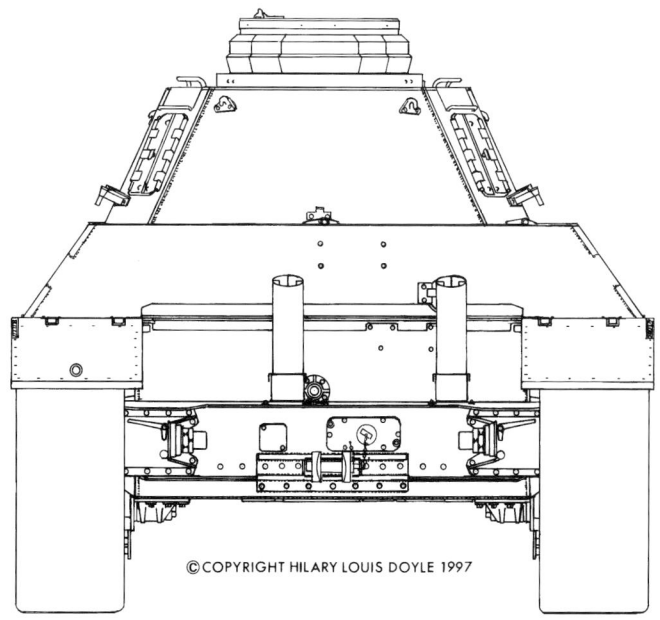

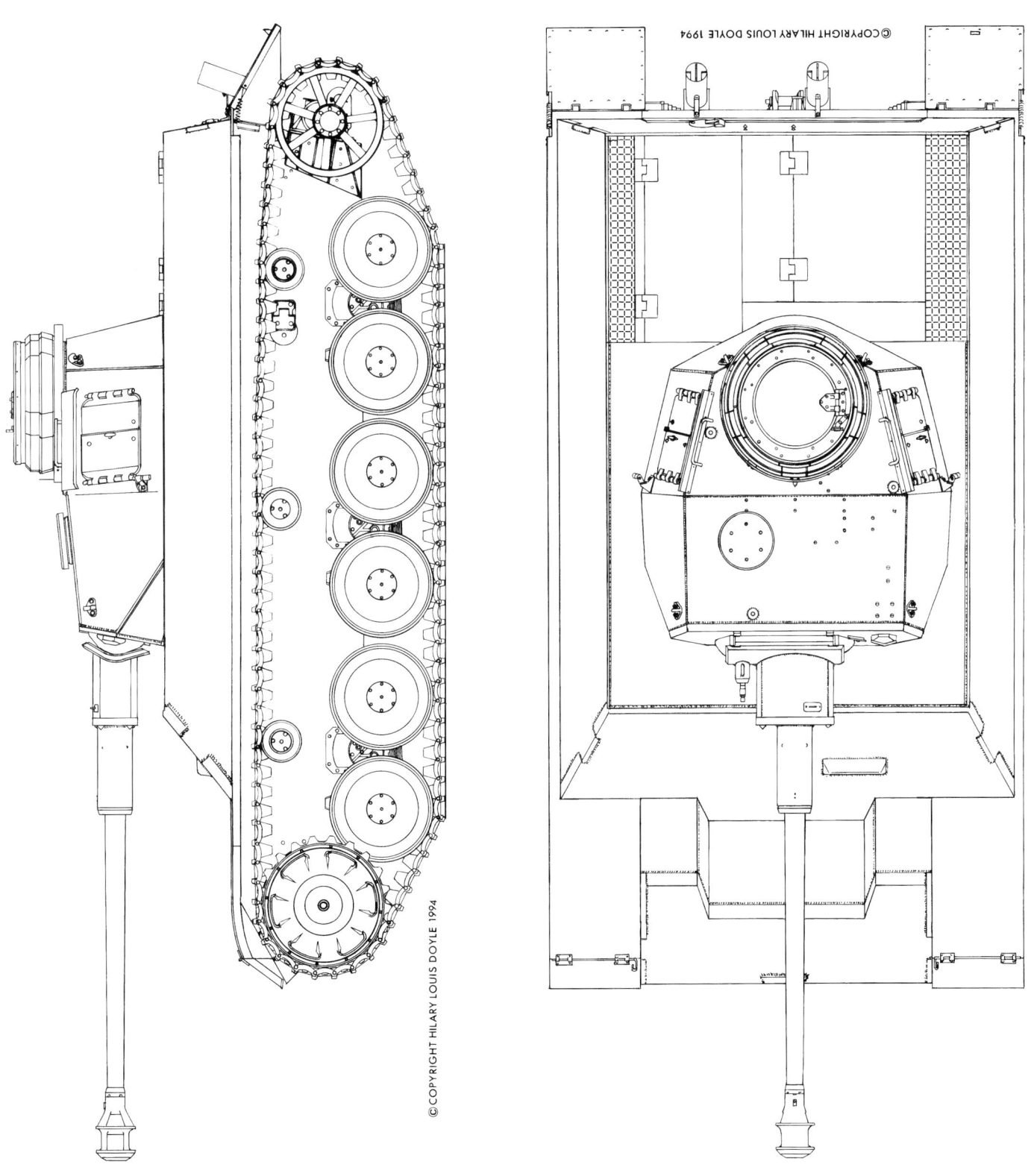

20-9

Vereinfachter Turm fuer Pz.Kpfw.IV

In drawing AKF 31941, Krupp proposed a "vereinfachten Turm (simplified turret) for the Pz.Kpfw.IV. The new turret design had a six-sided form, flat gun mantle, no vision or pistol ports, no commander's cupola, two hatches in the turret roof, and a simplified hatch in the left turret side.

Armor protection was significantly increased. A wider flat 80 mm plate replaced the curved 50 mm gun mantle. The thickness of the turret front plate was increased to 80 mm - still at 10 degrees. Side and rear plates were increased to 42 mm thick and sloped at 25 degrees from horizontal. The turret roof consisted of a single 25 mm thick plate sloped at 86 degrees.

A decision was made to drop this design on 19 July 1944. Plans had been made to completely stop Pz.Kpfw.IV production within the next few months and replace it entirely with Panzerjaeger mit L/70 (the Panzer IV/70 series).

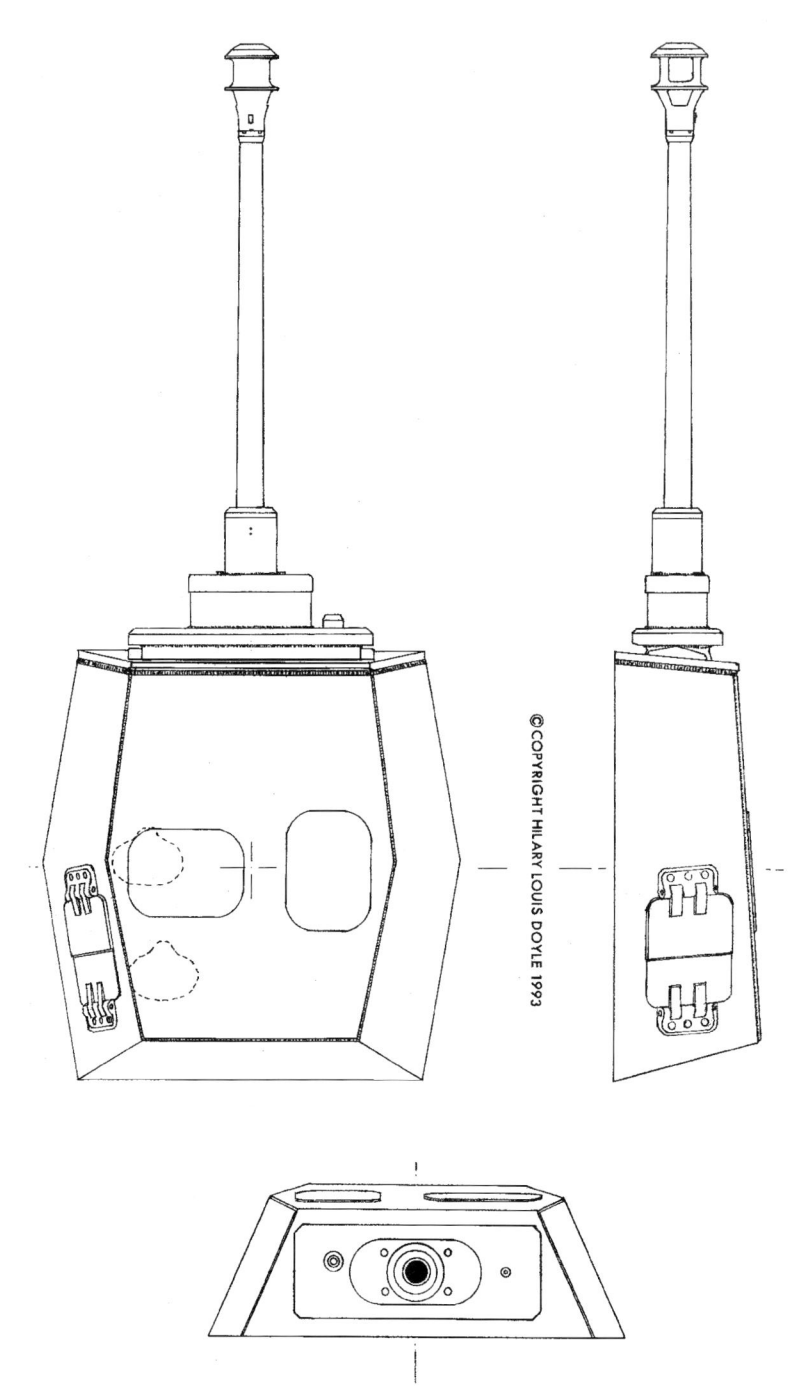

Mehrzweckpanzer

In June 1943, General-Oberst Guderian specified that a new standardized Mehrzweckpanzer (multipurpose tank) be developed for Aufklaerung (reconnaissance), Artillerie-Beobachtung (artillery observation), Flakpanzer (anti-aircraft tank), leichte Panzer-Jaeger (light tank destroyer), leichte Selbstfahrlafette (light self-propelled carriage) and other purposes on the basis of a 28 ton vehicle. Plans had been made to cancel Pz.Kpfw.IV production, and the heavy Tiger and Panther chassis were not to be diverted for these functions.

Wa Pruef 6 selected Krupp to develop the detailed design of this Mehrzweckpanzer, also known as the VK 28.01. In early June 1943, Krupp initially proposed a suspension with six 700 mm diameter roadwheels rolling on 600 mm wide tracks with which a 26 metric ton vehicle could maintain a sustained speed of 30 km/hr.

Among the variants proposed, Krupp completed a conceptual design of a Mehrzweckfahrzeug mit BW-Turm (multipurpose vehicle with Pz.Kpfw. IV turret) on 20 July 1943. While the vehicle still had a conventional front sprocket drive, the drive train selection was unusual with an Argus model 12LD330H rated at 550 horsepower and an Olvar 55 11 17 transmission and steering unit from the Leopard project. Overall chassis length was 5.685 m, width with Schuerzen 3.220 m, and height 2.595 m. With a 2.530 m wide wheelbase and 3.675 m track contact length, the steering ratio was 1.42.

Well-sloped armor was in fashion with the 50 mm thick upper hull front at 55 degrees, 50 mm lower hull front at 45 degrees, 30 mm superstructure sides at 30 degrees, 30 mm lower hull sides at 0 degrees, 30 mm hull rear at 25 degrees, 20 mm deck, and 20 mm forward belly plate, reduced to 16 aft.

The turret was standard from a Pz.Kpfw. IV Ausf.G with 50 mm front, 30 mm sides and rear (refer to Panzer Tracts No.4 for additional details). The drawing was made using a 7.5 cm Kw.K. L/43, even though the L/48 had come into production before this time. A unique feature was the cast armor commander's cupola with periscopes (similar in design to a Panther's). This cupola design was proposed for the standard Pz.Kpfw.IV series but was dropped based on a decision to stop Pz.Kpfw.IV production in the Fall of 1944.

In July 1943, the development plan envisaged taking 3 months to complete the detailed design, 6 months to complete a trial vehicle, 3 months to test the trial vehicle, and 8 months to perfect the design, so that series production could begin in April 1945.

At the end of October 1943, Krupp was informed by the Panzeroffizier under the Chef Gen. St.d.H. that a decision had been made to cease development of the Mehrzweckpanzer.

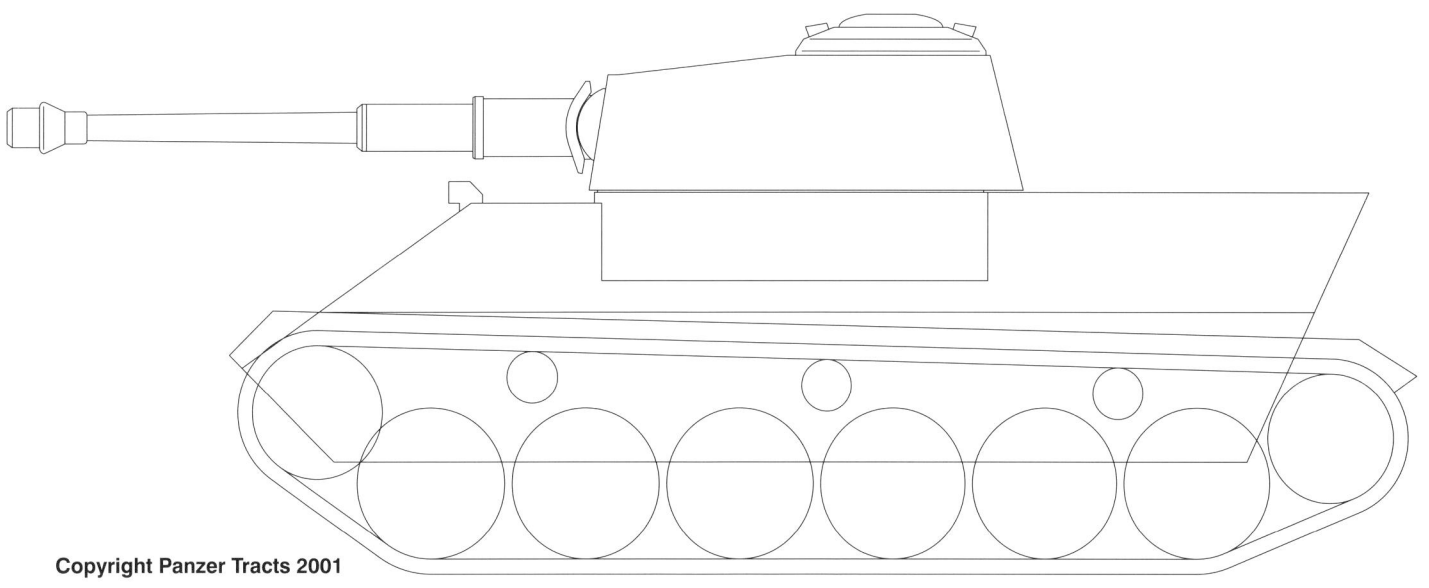

Copyright Panzer Tracts 2001

Porsche Typ 245-010
Leichter Panzerkampfwagen
mit 5.5 cm vollautomatischer Waffe

The necessity for developing a completely new type of tank was raised at a joint meeting of the Panzer- und Waffenkommission on 27 May 1943. Hand-held anti-tank weapons had been developed for the infantry (capable of penetrating over 110 mm armor at 400 meters range) and it was reported that the enemy had developed a new type of bomb scattering 1 kg hollow charge bomblets capable of penetrating 40 to 50 mm roof armor. It was concluded that the current design for tanks had reached a dead end and significant advances weren't expected. Therefore they had to start designing a completely new type of tank immediately that was especially proof against aircraft bombs (pointed shape). Military specifications for this vehicle were to be provided to development firms by 1 July 1943. The firms were to start tackling the basic problem now in order to obtain new workable proposals that were free from past practice. New blood was needed, because current design shops such as Henschel and M.A.N. were in a rut. The use of cast armor was to play a special role in the new tank shapes.

The second topic at this meeting was anti-aircraft defense for tanks. At least 18 Flak-Panzer, armed with a 2 cm quad, 3.7 cm double or triple, or 5.5 cm single Flak guns, would be needed for the defense of a Panzer-Regiment. A Flak-Panzer weighing about 30 tons with a sufficiently powerful engine to achieve a maximum speed of 50 to 60 km/hr was considered adequate for this role.

The team of Porsche and Rheinmetall tackled this problem - Porsche designing the chassis and Rheinmetall the weapons system - with incredibly unique solutions. One of their proposed solutions was the Porsche Typ 245-010 Leichter Panzerkampfwagen zur Verwendung gegen Erd- und Luftziele mit 5.5 cm vollautomatischer Waffe (light tank for use against ground and air targets with a 5.5 cm fully automatic weapon) dated 2 August 1943. 350 rounds, carried in metal belts, fed the Rheinmetall 5.5 cm automatischen Kanone MK 112 mounted in a turret with all-round traverse and an elevation arc from -8 to +82 degrees. Muzzle velocity was 600 m/s when firing 1.4 kg rounds. A 7.92 mm machine gun was mounted in the turret but in a ball mount with independent traverse and elevation instead of a normal coaxial configuration. For close defense, a grenade thrower with adjustable elevation was mounted in the rear turret roof.

The gunner had a periscopic sight protected by an armor housing on the turret roof hatch and the commander had a pivoting (all-round) periscope mounted in his turret roof hatch. The driver had a port for direct vision and a single periscope for use when buttoned up. An escape hatch in the left superstructure side was provided for the driver.

The main bodies for the hull and turret were to be cast armor. Hull armor protection consisted of 60 mm glacis at 55 degrees, 60 mm hull front at 30 degrees and rounded, 40 mm superstructure sides at 28 degrees, 16 mm plus 16 mm hull sides at 0 degrees, 25 mm rear engine deck at 62 degrees, 25 mm hull rear at 25 degrees and rounded, and 16 mm roof and belly. The turret armor was 40 mm thick all round at 30 degrees with 20 mm on the roof.

The drive train consisted of an air-cooled V-10 Porsche Typ 101 engine rated at 345 horsepower with a hydraulic transmission and steering unit. The entire drive train was mounted at the rear for the rear drive sprockets. Designed for a maximum speed of 65 km/hr, it carried sufficient gasoline for a range of 240 kilometers by road or 150 cross-country.

Six 600 mm diameter steel-tired roadwheels were mounted in pairs with a vertical volute spring suspension. A track width of 480mm and ground contact length of 2.920 meters resulted in a ground pressure of 0.64 kilogram/square centimeter for the 18 metric ton Panzer. A wheelbase of 2.320 meters resulted in an acceptable steering ratio of 1.25. Overall length was 4.880 meters, width 2.320 meters, height 2.575 meters, and ground clearance 0.445 meter.

Similar conceptual designs were proposed by Porsche and Rheinmetall from July 1943 to January 1944 as multipurpose anti-aircraft and reconnaissance Panzers using the same basic chassis from Porsche but with the Rheinmetall 5.5 cm MK 112 mounted in the hull. No evidence has been found that production of a single trial vehicle was ordered or that serious consideration was given to their eventual series production.

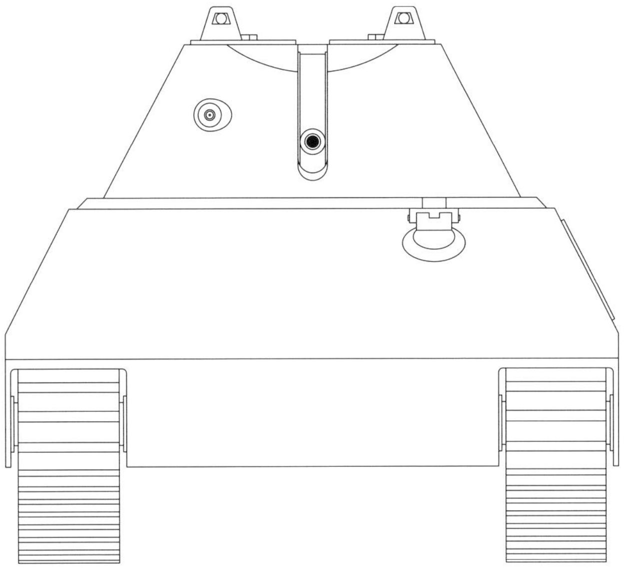

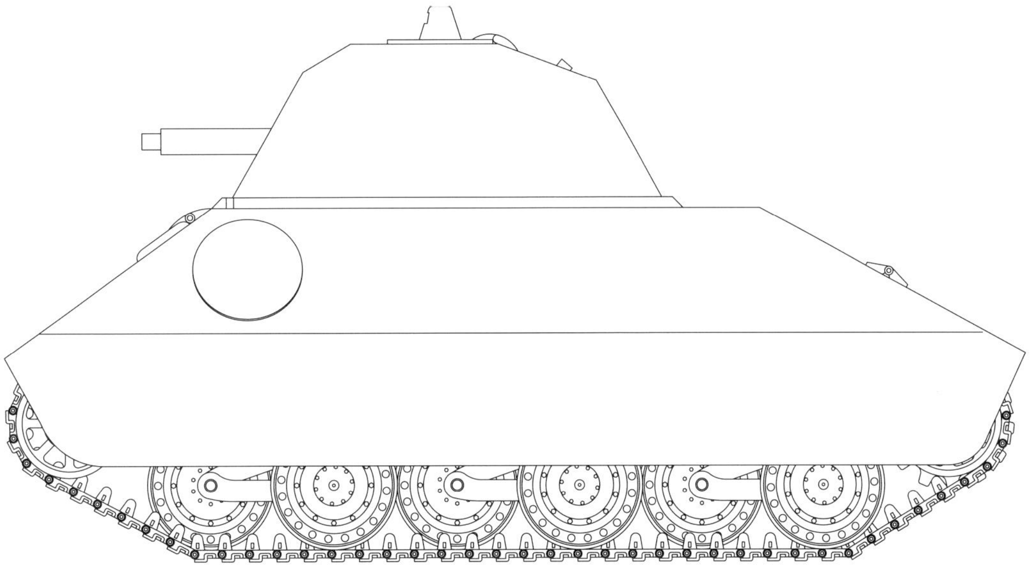

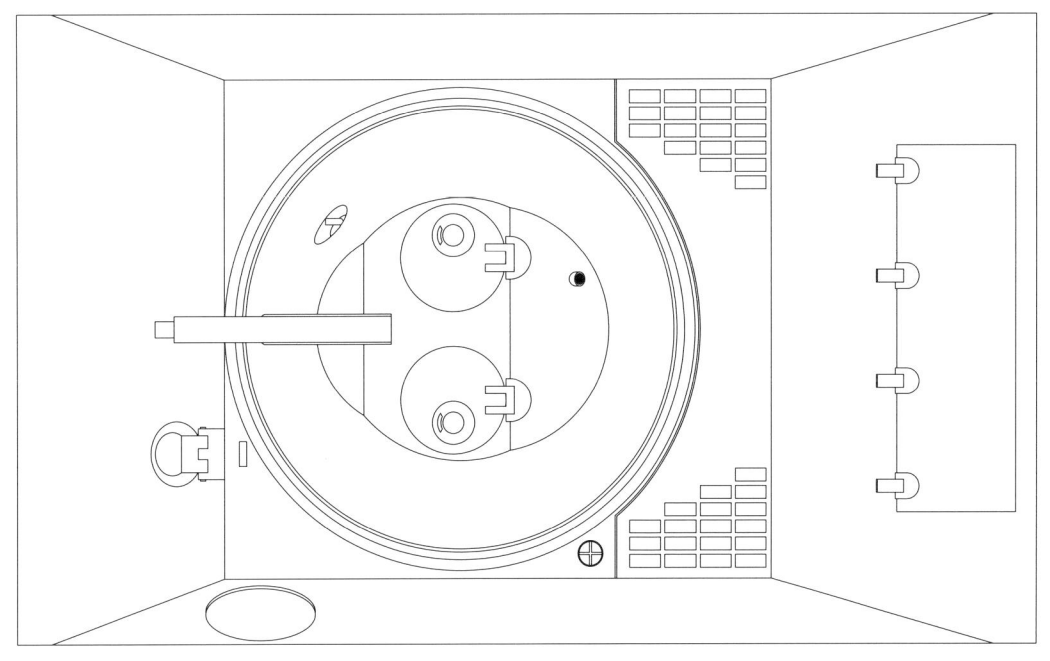

The wooden model of a Porsche Typ 245 le.Pz.Kpfw.mit 5.5 cm Waffe (WJS)

Panzerkampfwagen VI H Ausf.H2

An alternative turret was to be designed for the VK 45.01, as Oberst Fichtner, head of Wa Pruef 6, reported on 27 September 1942: In a meeting on 25 July 1941 in Stuttgart, I informed Prof.Dr. Porsche that I was not happy with the Krupp turret and strived for a better solution for the future that would be equally suitable for both the Pz.Kpfw.Typ Porsche and Typ Henschel. As already reported to Minister Dr. Todt, Wa Pruef 6 gave Rheinmetall a contract in mid-July 1941 to design a turret with a gun that could penetrate 140 mm thick armor at a range of 1000 meters without specifying that the caliber had to be 8.8 cm. The authority for this comes from Hitler's directive dated 26 May 1941, which states: "If the same penetration capability can be achieved by a smaller caliber than the 8.8 cm (i.e., 6 or 7.5), this should be given preference based on increased ammunition load and the lower turret weight. The chosen caliber must be suitable for engaging tanks, ground targets, and bunkers." Rheinmetall is attempting to achieve the penetration ability with a normal cylindrical gun tube based on the same principles as the Pak 44. It would have been wrong to pass up this Rheinmetall project because in meeting all the necessary requirements (rate of fire, ammunition load, balanced turret weight, observation conditions), it foreseeably leads to a more advantageous turret than the current Krupp turret.

The first 7.5 cm Kw.K. Versuchsrohr (trial gun tube) of L/60 caliber length, designed and test fired by Rheinmetall-Borsig, just met the specified penetration ability of 100 mm thick armor plate at 30 degrees at 1400 meters range. Therefore to ensure that the penetration specification was met, a final gun tube of L/70 caliber length was chosen. By 11 February 1942, Rheinmetall had designed the VK 45.01 (Rh) turret with a 7.5 cm Kw.K.42 to be mounted on the VK 45.01 (H) chassis. The VK 45.01 (H) with Rheinmetall-Turm mit 7.5 cm Kw.K. L/70 had been officially designated as the Pz.Kpfw. VI H Ausf.H2 by Wa Pruef 6 by 1 July 1942.

On 1 July 1942, Wa J Rue (WuG 6) revealed long-range plans under Hitler Panzerprogramm II to produce only the first 100 production series VK 45.01 (H) with the 8.8 cm Kw.K. L/56. Then starting with the 101st VK 45.01 (H) in February 1943, production was to be shifted to the Rheinmetall turret with the 7.5 cm Kw.K. L/70.

The subject of Tiger armament was discussed at a Panzerkommission meeting on 14 July 1942: Recently the ability to penetrate 100 mm of armor was also achieved with the 8.8 cm Kw.K. L/56; therefore conversion to the 7.5 cm Kw.K. L/70 is no longer necessary. Conversion to the 8.8 cm Kw.K. L/71 should occur at the end of this year. This decision resulted in the entire VK 45.01 (H) production run being outfitted with turrets mounting the 8.8 cm Kw.K. L/56.

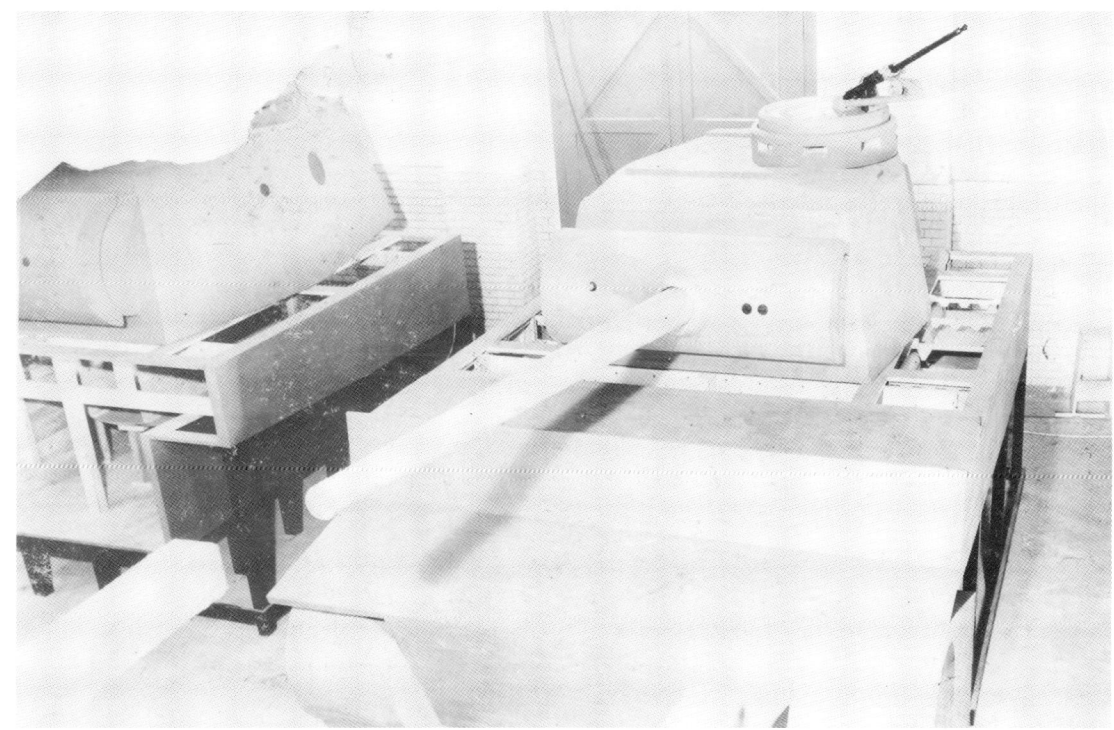

The wooden models of the Rheinmetall-Turm mit 7.5 cm Kw.K. L/70 for the Pz.Kpfw.VI H Ausf. H2 with pistol ports on the sides, an emergency escape hatch on the right side, a communication port on the left side, and a machine-gun ball mount on the rear (APG)

8.8 cm Kw.K.43 L/71 in a modified Panther-Schmalturm

In a meeting of the Entwicklungskommission Panzer on 23 January 1945, Oberst Holzhaeuer (Wa Pruef 6) reported that development of a Panther with the 8.8 cm Kw.K. L/71 in a Panther-Schmalturm (narrow turret) was to be accomplished by Daimler-Benz. The turret ring diameter was to be 100 mm larger than the current Panther turret, with an increase in weight of about 1 metric ton. Ammunition stowage amounted to 56 rounds in comparison to 103 rounds previously stowed in the Panther. A wooden model had been completed.

Krupp had previously created a sketch of an 8.8 cm Kw.K.43 L/71 in a Panther-Schmalturm that had been modified as little as possible (drawing number Hln-130 dated 18 October 1944). Krupp was awarded a development contract by Wa Pruef 6 on 4 December 1944. In a letter to Wa Pruef 6 dated 12 February 1945, Krupp explained that as a basis for their proposal, whenever possible, the Panther-Schmalturm with accessories had been left unchanged. The 8.8 cm Kw.K.43 gun could be installed if the trunnions on the carriage were moved further back 350 mm (i.e., the gun moved forward 350 mm).

In February 1945, Wa Pruef 6 decided to have Daimler-Benz further develop the Panther-Schmalturm with 8.8 cm Kw.K.43. Only an experimental turret fabricated from soft steel was to be completed. The design specifications were:
o Elevation from minus 8 through plus 15 degrees.
o Only the 8.8 cm Kw.K.43 previously developed for the Tiger II was acceptable. The recoil and return cylinders were to be mounted above the gun with the bore evacuation cylinder in the middle. The muzzle brake was to be dropped and the trunnions were to be relocated.
o A smooth armor plate for the turret front with the smallest possible apertures for the main gun and machinegun. The middle of the trunnions were to be located on the forward edge of the front plate.
o A rangefinder was to be included. An attempt was to be made to use the already available 1.32 or 1.65 meter rangefinders designed for Panzerkampfwagen.
o Special value was placed on a low turret height.
o The free turret ring diameter was to be 1750 mm to provide the loader the necessary room to maneuver.
o Ammunition had to be easily accessible in ready racks in the turret.
o The commander's cupola and turret traverse gear were to be the same as in the current Panther-Schmalturm.
o Consideration was to be given to mounting the S.Z.F.2 or S.Z.F.3 stabilized gunsights.
o The rear wall of the turret was to be sloped, instead of upright as was the case in the first wooden model from Daimler-Benz.

On 8 March 1945, Oberst Crohn (Wa Pruef 6) requested that Krupp complete a design for the armor shell of a Panther Ausf.F turret mounting an 8.8 cm Kw.K.43 by 12 March 1945.

On 14 March 1945, further development of the Panther was discussed with the Generalinspekteur der Panzertruppen. A new situation had been presented as a result of the excellent work by the Waffenamt in designing an 8.8 cm Kw.K. L/71 (Tiger II gun) in a Panther. 15 main gun rounds were accessible in the turret along with about 50-54 rounds stored in the hull. With a rangefinder protected by armor and a gunsight with a stabilized view, it was about the same as the Panther-Schmalturm. Weight was about one metric ton heavier than the current Panther. Wa Pruef 6 was to be especially thanked for development of this Panther. If production of the "8.8 cm Panther" was successfully started, Wa Pruef 6 was to make preparations for the future to convert all available Panthers that underwent major overhaul to mounting an 8.8 cm turret. The Versuchs-Panther in soft steel was to be completed by early June. If the necessary support was provided, series production was to start in the last quarter of 1945.

On 14 March 1945, the Generalinspekteur der Panzertruppen requested that Wa Pruef 6 provide a Versuchs-Panther with an 8.8 cm Kw.K. L/71 completed in accordance with the wooden model from Daimler-Benz that had been displayed on 12 December 1944. The Generalinspekteur der Panzertruppen agreed to a normal Panther hull with a modified superstructure and turret in soft steel. Wa Pruef 6 was requested to expedite completion and to ensure the timely display of the Versuchs-Panther.

On 23 March 1945, Speer relayed the request that Hitler wanted a Panther with an 8.8 cm Kw.K. be displayed about mid-April 1945 along with other weapons. When interrogated after the war, representatives from Daimler-Benz stated that plans had been made to eventually mount the 8,8 cm Kw.K.43 L/71 with a stabilized gunsight in the Schmalturm, but this project was not far advanced. In August 1945, a wooden mock-up was still located at the Daimler-Benz assembly plant.

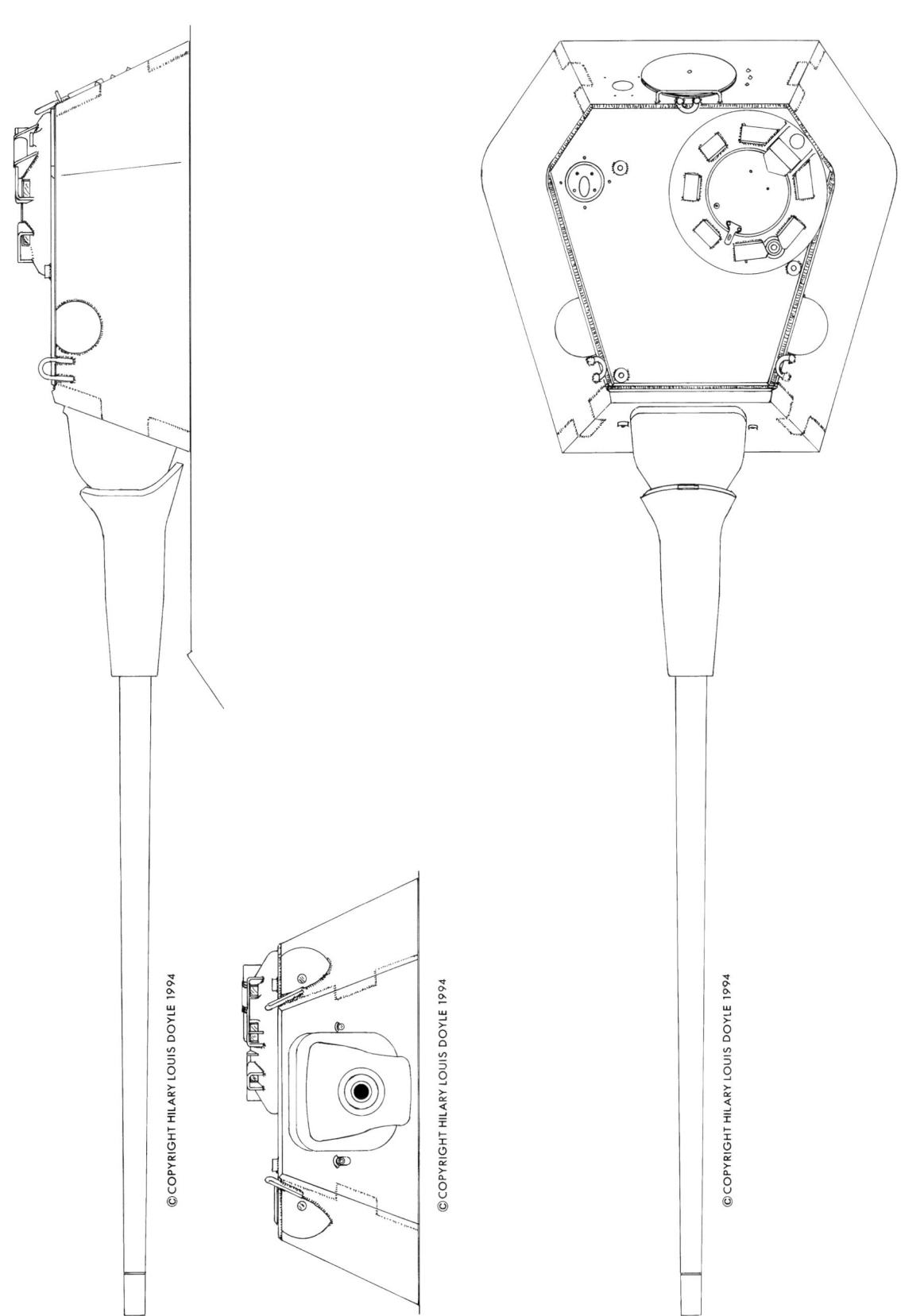

E 50/E 75

As related postwar by Oberbaurat Kniepkamp (civilian head of automotive design in Wa Pruef 6): The E-Serie program was conceived in May 1942 and authorized as a project in April 1943. Several new principles were to be introduced with this new series: 1) To achieve a very strong front plate, move all possible weight to the rear, 2) unify the drive train unit to simplify maintenance and service, 3) standardize all Panzers into four weight classes (E 10, E 25, E 50, and E 100), 4) attach all suspensions from the outside and no fighting space encumbered by through torsion bars, and 5) in case the front idler or any roadwheels were destroyed by mines, the vehicle must be capable of proceeding by adjusting the track around the remaining wheels. The technical superiority of forward drive was recognized, but the military advantage of having the drive at the rear where it was not endangered by mines or gunfire influenced the choice of rear drive.

Adlerwerke was chosen for the detailed design of the suspension for the E 50/E 75. As related postwar by Dir. Jenschke of Adlerwerke: The E 50 and E 75 were planned as so-called standard tanks. They were to have the same engines, rear drives, tracks, idler wheels, track tensioning adjustor, and ventilation system. Fuel tanks and other accessories were to be identical. The hull structures were to have the same shape and dimensions. It was planned to keep the same outer dimensions for the both hulls and give E 50 the advantage of an increased interior which would come about by the use of a thinner rolled armor plate for the lighter Panzer. This project enabled uniform handling of the two Panzers, i.e., they could be worked on with the same tools and could be constructed on the same assembly line.

A new engine was planned which was an improved Maybach HL 230, retaining the same swept volume. The engine's reliability was improved by strengthening the crankshaft bearing and connecting rods. Power was increased by adding a supercharging compressor and a direct fuel injection system. However, the supercharger was not yet developed, and it was proposed to install the Maybach HL 234 with fuel injection which was rated at 900 metric horsepower at 3000 rpm.

Up to this design, the transmission, steering units, and two final drives had been fitted as separate components. These four separate units were designed to fit into one casing as the rear drive for the E 50 and E 75, so that final drives bolted onto the hull side disappeared. The transmission consisted of a hydraulic pre-selected 8-speed gearbox combined with a dual radius steering unit. Maximum speed was to be 60 km/hr for the E 50 and 40 km/hr for the E 75. This could be achieved by just changing one gear on each side of the otherwise standardized drive train.

The suspension was an entirely new design. The individual roadwheels were mounted in Schritt-anordnung (step arrangement). The crank arms for each pair of roadwheels acted through ratchets and wedge gear segments on a large parcel of Belleville springs contained in a housing with the shock absorbers. The E 50 had three suspension units on each side, and the E 75 had four suspension units. The combat track for the E 50 was to be used as the transport track for the E 75.

Alternative drive train units were also being developed, including an 8-cylinder Klockner-Humboldt-Deutz diesel engine, a Voith torque converter, and a Mech-Hydro transmission. No actual construction or assembly work had been completed by the end of the war, only development of components. Nor had any plans for complete assembly been formulated.

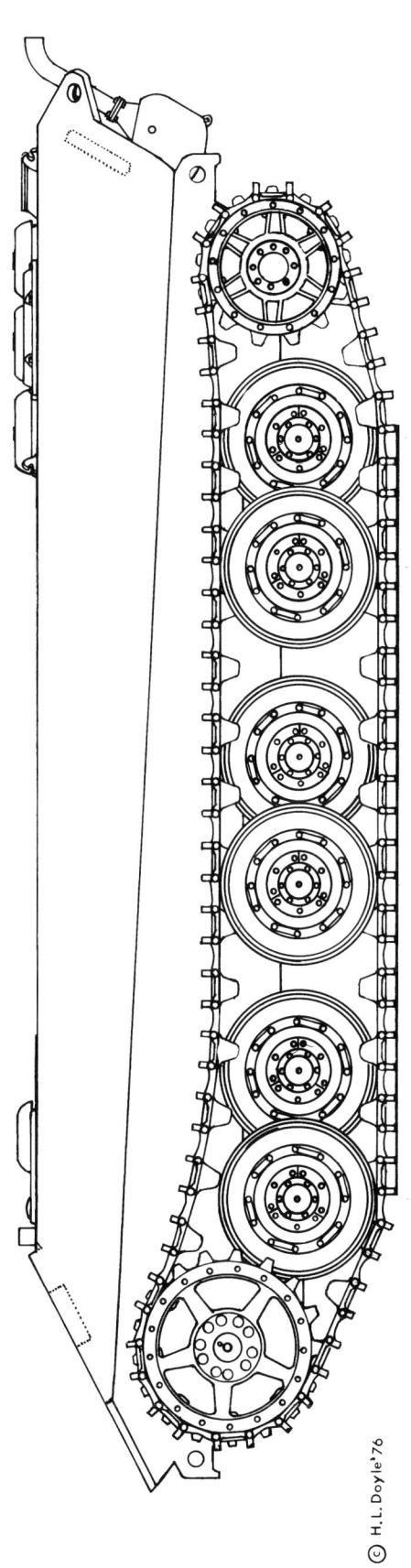
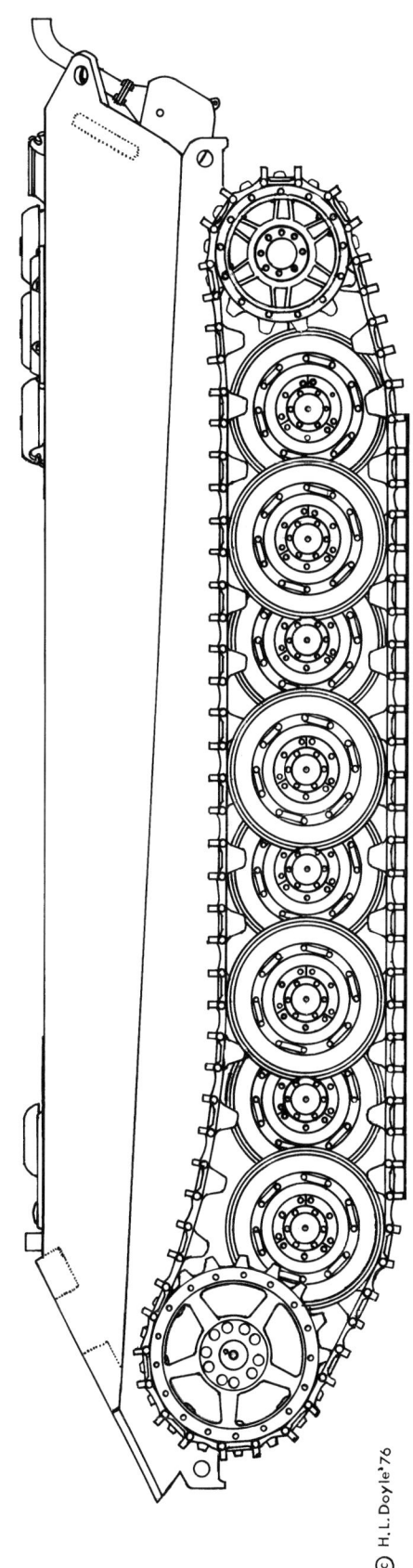

20-19

Panzerkampfwagen "Tiger" Ausf.P2
Fgst.Nr. Serie 150101 - 150300
Porsche Typ 180

On 26 May 1941, Hitler requested that an 8.8 cm gun firing an armor-piercing projectile capable of penetrating 100 mm thick armor plate at a range of 1500 meters be mounted in a turret on the Porsche chassis. A new turret was designed by Krupp, with the higher performance 8.8 cm gun and its longer ammunition, longer recoil, and longer gun needing counterbalancing. Plans were made to mount 8.8 cm Kw.K. L/71 guns in turrets starting with the 101st Pz.Kpfw.VI (VK 45.01 (P)). The model designation was changed from Typ 101 verstaerkt (type 101 strengthened) to Typ 180 and from VK 45.01 (P2) to VK 45.02 (P) on 23 March 1942.

The initial concept for the new Panzer was that the chassis designed for the VK 45.01 (P) could be adopted with very little modification aside from that needed to fit the new turret and ammunition stowage. Starting in January 1942, the protection afforded by the armor was greatly increased by sloping the plates. This resulted in modifying the driver's hatch, driver's visor, and machinegun ball mount to fit the sloped glacis plate. The 80 mm glacis plate was sloped at an angle of 55 degrees, 80 mm front nose plate at 55 degrees, 80 mm superstructure side plates at 15 degrees, 80 mm hull side plates at 0 degrees vertical, 80 mm upper tail plate at 60 degrees, 80 mm lower tail plate at 25 degrees, 40-25 mm deck plates at 90 degrees horizontal, and 40 mm front and 20 mm rear belly plates at 90 degrees horizontal.

The drive train consisted of twin Porsche Typ 101/3, 10-cylinder engines connected to electric generators providing power to two electric motors, one for each drive sprocket at the rear of the hull, designed to provide a maximum speed of 35 kilometers per hour. The combat weight of 45 metric tons was distributed over three sets of two 700 mm diameter steel-tired roadwheels per side suspended by lateral torsion bars. The unlubricated 640 mm wide tracks provided an acceptable ground pressure (when the tracks sank to 20 cm) of 1.06 kilograms per centimeter squared.

Contracts for fabrication of the components and assembly of a production series of 100 were awarded in February 1942, with the first six VK 45.01 (P2) with 8.8 cm Kw.K. L/71 guns to be completed in November 1942. Another 100 were ordered in April 1942. Because of problems with the Porsche-designed engines and suspension, all contracts for the production series were canceled on 3 November 1942. Wa Pruef 6 issued new contracts to Krupp for three operational turrets and three armor hulls for Pz.Kpfw.Tiger P2 in January 1943. In February 1943, Prof.Dr. Porsche reported that three Tiger 2 (VK 45.02 (P)) with electrical drive and new suspensions were being completed at Nibelungenwerk.

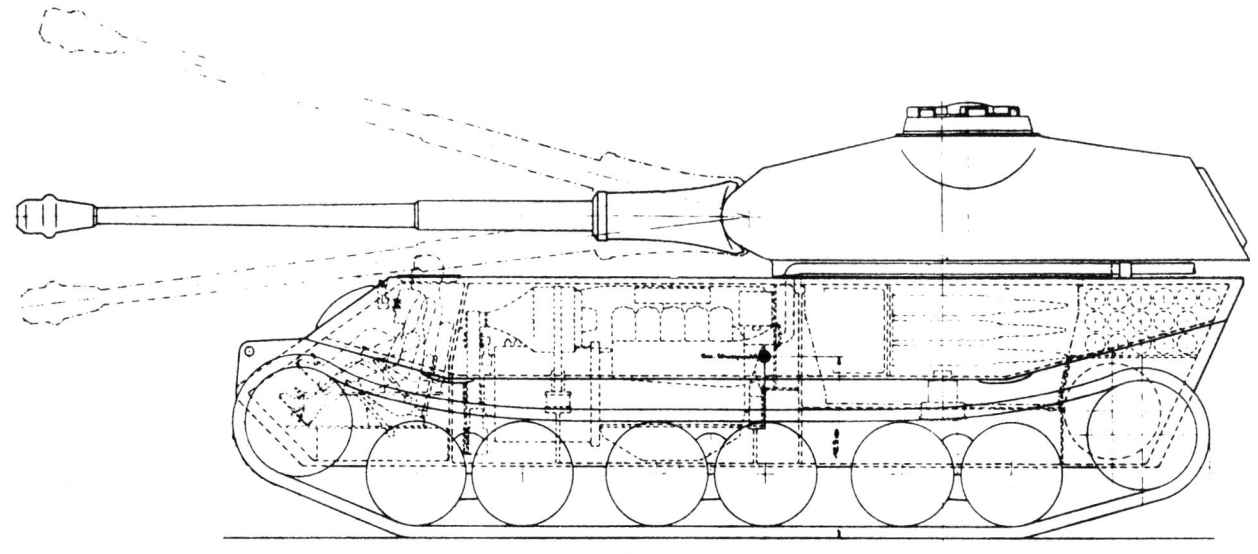

The Porsche design for the Typ 180 with the turret mounted at the rear and the twin V-10 engines mounted in the center

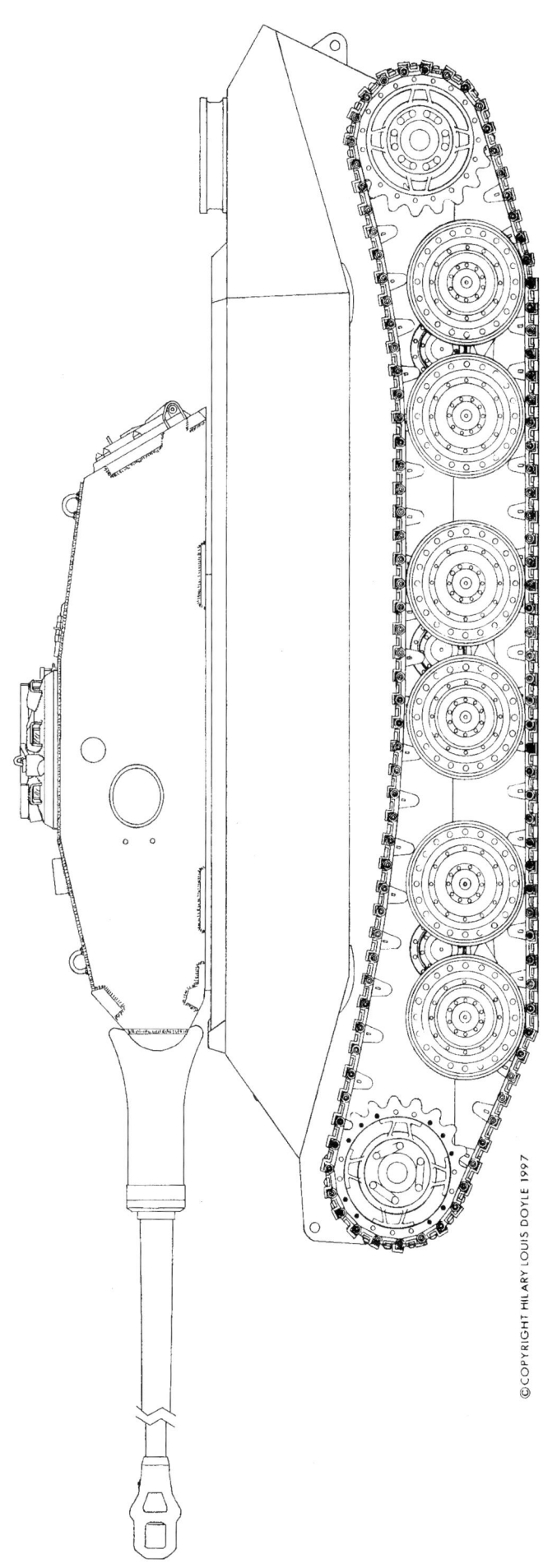

Panzerkampfwagen VII
(VK 65.01)

At a meeting on 19 January 1939, Wa Pruef 6 revealed that they were developing Panzers weighing 30 and 65 metric tons with a 7.5 cm Kw.K. in the turret and the same interior space as the B.W. Heavier armament was rejected in order to achieve sufficient armor protection. The 65 ton Panzer was to have 80 mm thick armor, proof against penetration by 5 cm Pzgr. A 10 cm gun had longer range but was less effective against bunkers. In addition, the longer and heavier 10 cm rounds were more difficult to stow and load. The highest speed required for the 65 ton Panzer was 20 to 25 km/hr.

In January 1939, Krupp was ordered to complete a full-scale wooden model of the S.W. turret with three different guns and mantlets; 7.5 cm Kw.K. L/24 (6.8 kg Pzgr. at 398 m/s), 7.5 cm Kw.K. L/40 (6.8 kg Pzgr. at 685 m/s), and 10.5 cm Kw.K. L/20 (15 kg shell at 420 m/s). The wooden model was completed in April 1939 and Krupp was ordered to complete a soft-steel Versuchsturm (trial turret) with hydraulic traverse. At the end of June 1939, Wa Pruef 6 made the final decision to mount the 7.5 cm Kw.K. L/24 in an S.W. turret with seven periscopes in a traversable commander's cupola. A decision was made in March 1940 that with the exception of armor thickness, the S.W. turret was to be an exact copy of the D.W. turret design.

Since it didn't appear possible to meet the railway profile restrictions with a single-piece hull, Henschel designed the hull in three pieces, which they estimated would take about three weeks to re-assemble when moved by rail. A Maybach HL 224 engine rated at 600 metric horsepower was to propel the 65 ton Panzer at a maximum speed of 20 km/hr. The nine pairs of rubber-tired roadwheels on each side were interleaved, were sprung by torsion bars, and ran on 800 mm wide tracks.

On 1 September 1939, In 6 ordered Wa Pruef 6 to produce a 0-Serie. Wa Pruef 6 awarded contracts to Krupp in February and March 1940 for the armor components for a 0-Serie of eight VK 65.01 and for turret assembly with guns in operational order. The first turret was scheduled to be completed in August 1942. Armor hulls and operational turrets were to be delivered to Henschel, which was to assemble the chassis and mount the turrets to complete final assembly. After the defeat of France, Wa Pruef 6 decided that Panzers weighing over 30 tons would not be tactically useful because of bridge weight restrictions. Contracts for the armor components were rescinded in August 1940, and work on the soft-steel S.W. Versuchsturm was halted in October 1940. Wa Pruef 6 reported that one soft-steel Versuchs-fahrgestell was being completed in mid-1941. Work on the VK 45.01 took priority, and at the end of 1942 Wa Pruef authorized Henschel to scrap the parts.

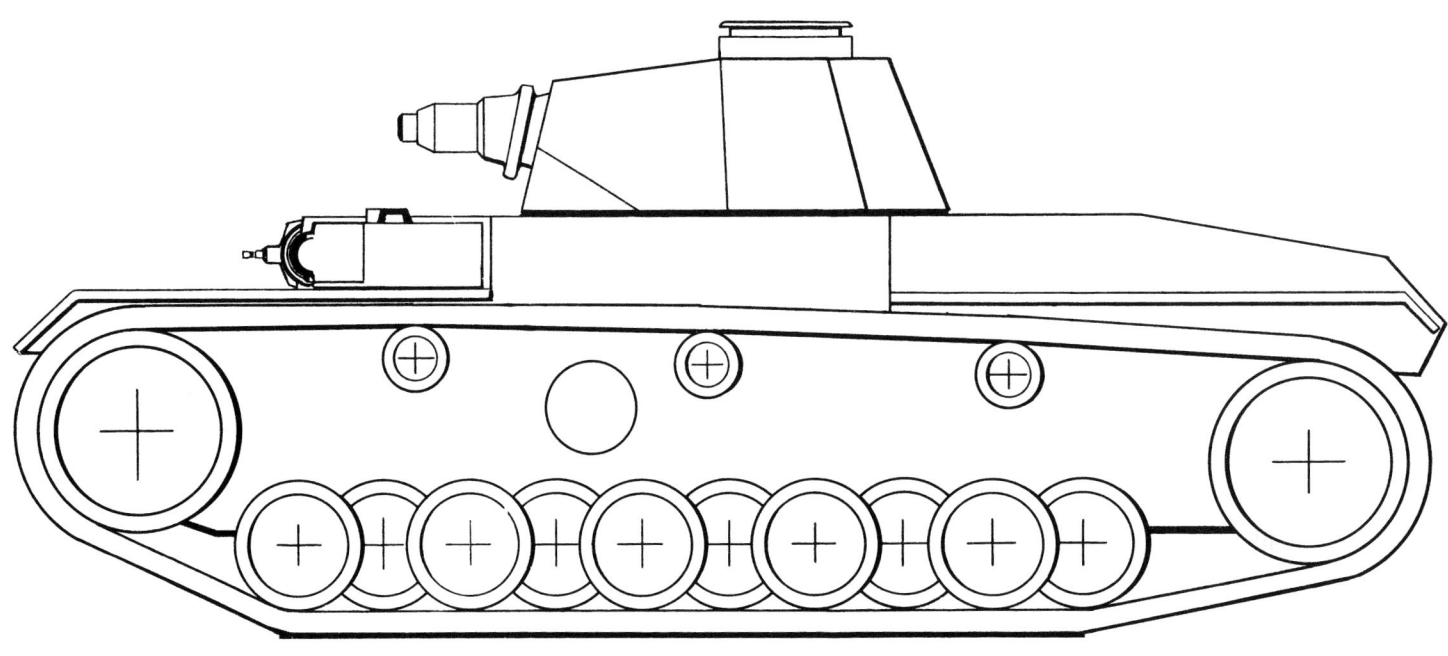

"Löwe"
Panzerkampfwagen VII
(VK 70.01)

On 1 November 1941, the general specifications were laid out for a new super heavy tank in the 70 metric ton weight class - the VK 70.01 with 140 mm thick frontal and 100 mm thick side armor. A maximum speed of 43.6 kilometers per hour was to be achieved by using a Daimler-Benz Schnellbootsmotor (torpedo boat engine) rated at 1000 horsepower at 2400 rpm. A crew of five were to man this Panzer, with three in the turret and two in the hull. The weapon wasn't specified, but the turret was to be fully traversable through 360 degrees.

Having lost out on the development of the VK 30 Series (Panther), on 17 December 1941 Krupp was asked to intensively pursue development of this Panzer so strongly armored that no known enemy anti-tank gun could penetrate it. An upper limit of 90 tons was considered the maximum that railroads could carry.

On 21 January 1942, Krupp presented their conceptual design of a VK 70.01 with a 10.5 cm Kw.K. L/70 (capable of penetrating a 160 mm thick armor plate at 30 degrees at 1000 meters) in the turret. Wa Pruef 6 informed Krupp that they were to use the new engine that Maybach was developing, the HL 230 engine rated at 800 metric horsepower, planned to go into series production in January 1943.

In February 1942, the war situation pushed Wa Pruef 6 into a decision to quickly complete two Versuchsfahrzeuge in the 72 ton weight class with the same drive train and armor as a Tiger (100 mm front, 80 mm sides) and to go into series production without testing. The 50 Tiefladewagen (low carriage rail cars) being produced for the 52 ton Tiger could carry this weight. Krupp was awarded contract SS 006-6307/42 to produce two Entwicklungsfahrzeuge VK 70.01 (one with a turret, the second with only a test weight). The name of the project was changed in April 1942 from VK 70.01 to Pz.Kpfw.Löwe (Lion).

Many alternative designs were considered from February to May 1942, including rear and center mounted turrets/engine compartment, different drive trail components, varying armor/weight, and alternative armament (Refer to Table on Evolution).

Following Hitler's decision to produce even heavier tanks, the contracts for development and production of the chassis were rescinded on 18 May 1942. On 20 July 1942, Wa Pruef 6 ordered Krupp to stop work on the Pz.Kpfw.Löwe turret.

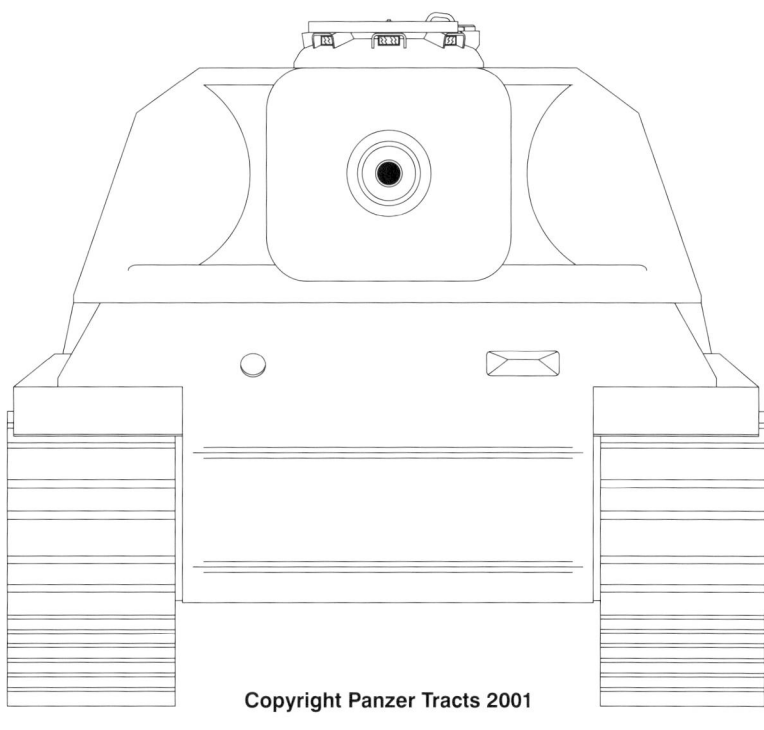

Copyright Panzer Tracts 2001

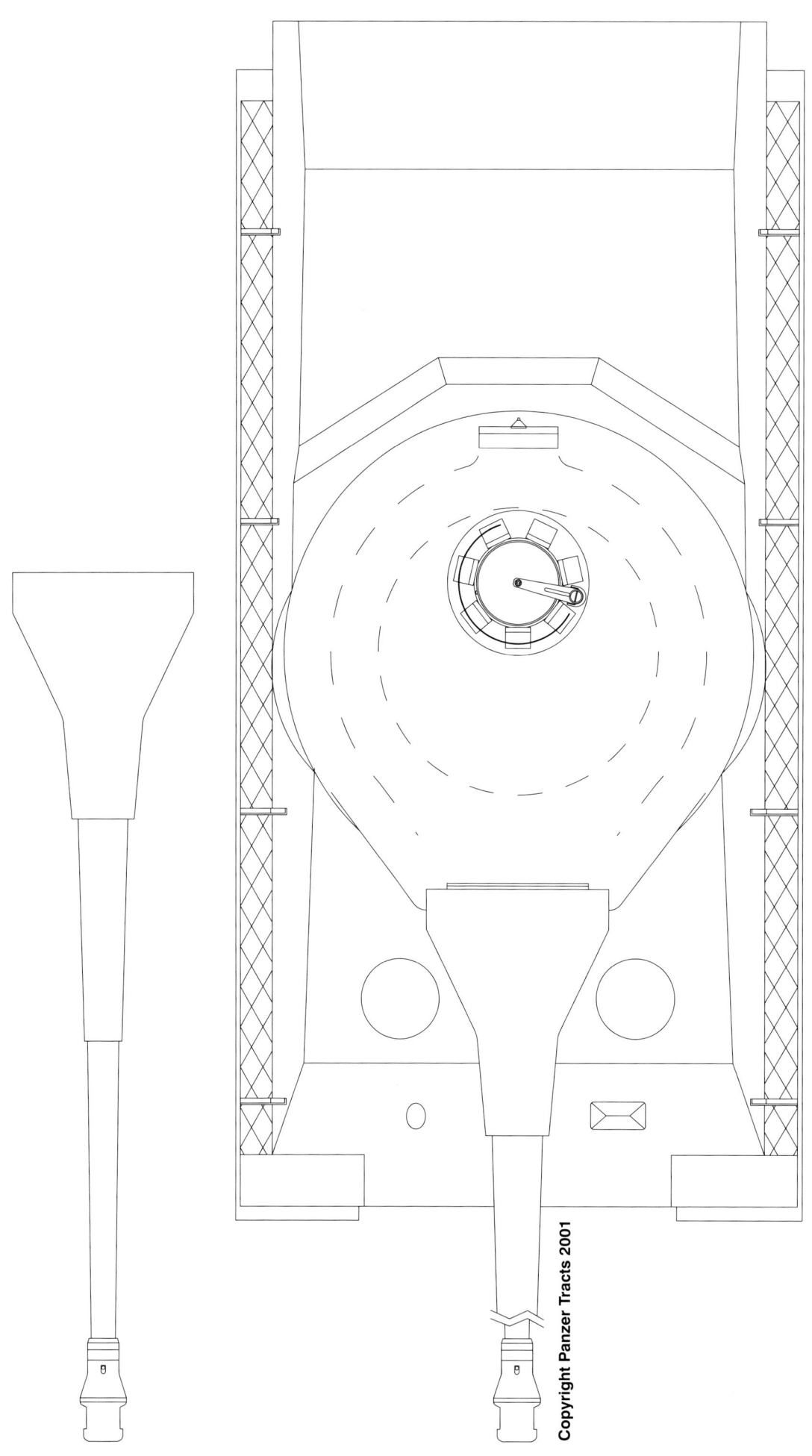

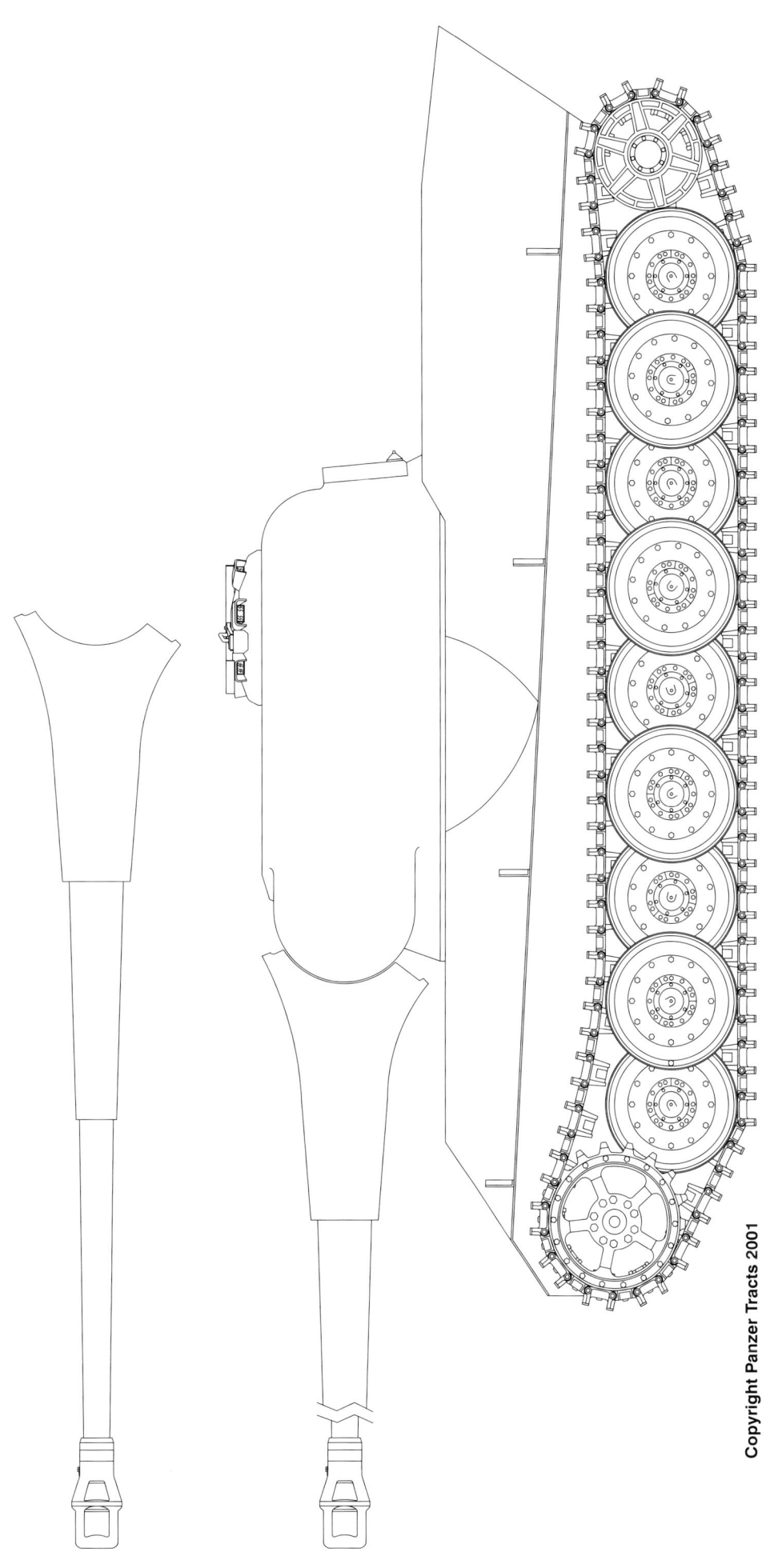

Krupp's conceptual design for a Panzerkampfwagen "Loewe" (Lion)

Evolution of VK 70.01/"Loewe" Conceptual Design

Name		VK 70.01	VK 70.01	VK 70.01	"Loewe"	"Loewe"	"Loewe"
Drawing No.				W 1661	W 1662	W 1663	
Date		27Feb42	9Apr42	9Apr42	23Apr42	23Apr42	11May42
Weight	t	75	76	90	90	80	90
Armament	cm	10.5 L/70	10.5 L/70	10.5 L/70	10.5 L/70	10.5 L/70	15 L/40
					or 15 L/40	or 15 L/40	
		1 M.G.	1 M.G.	1.M.G.	1 M.G.	1 M.G.	1 M.G.
Elevation	degrees	-8,+38	-8,+38	-8,+38	-8,+38	-8, +38	-7,+30
Ammunition		80	76	76	80 (10.5)	80 (10.5)	80
		2000 M.G.	2000 M.G.	2000 M.G.	2000 M.G.	2000 M.G.	2000 M.G.
Armor							
Hull Front	mm	100	100	120	120	100	120
Hull Side	mm	80	80	80/100	80/100	80	80/100
Hull Deck	mm				40		40
Turret Froont	mm	100	100	120	120	100	120
Turret Roof	mm	40	40	40	40	40	40
Total Length	mm	11250	11240	11640	11670	11220	10760
Chassis Length	mm	6970	7050	7450	7450	7450	7740
Total Width	mm	3540	3630	3630	3830(4030)	3830(4030)	3830(4030)
Total Height	mm	3030	2960	2960	3085	3085	3080
Firing Height	mm	2510	2470	2470	2495	2495	2480
Track Width	mm	750	800	800	900(1000)	900(1000)	900(1000)
Wheel Base	mm	2790	2830	2830	2930(3030)	2930(3030)	2930(3030)
Track Contact	mm	4340	4340	4800	4960	4340	4960
Ground Pressure	kg/cm2	1.12	1.10	1.17	1 (0.9)	1 (0.92)	1 (0.9)
Ground Clearance	mm	500	480	480	500	500	500
Maximum Speed	km/hr	30	26.8	23	35	35	30
Engine	HP/rpm	700/3000	700/3000	800/3000	HL 230	HL230	2 Porsche
Transmission		12 EV 170	12 EV 170	12 EV 170	12 EV 170	12 EV 170	
Steering		L600C	L600C	L600C	L600C	L600C	KLL 800
Suspension		Torsion bar	Torsion bar	Torsion bar	Torsion bar	Torsion bar	Torsion bar

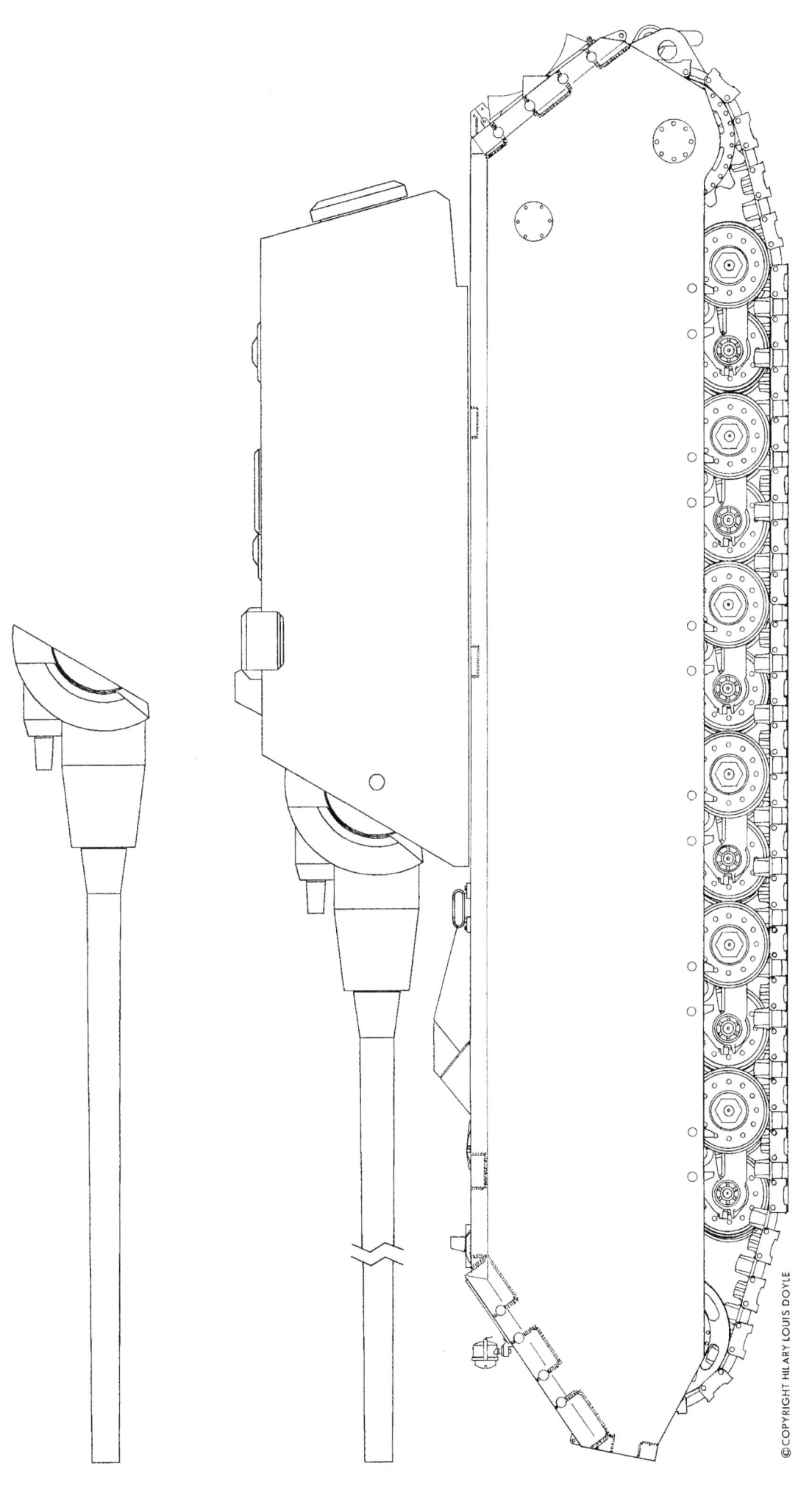

Panzerkampfwagen "Maus II"

A conceptual sketch of a Maus II turret with a larger turret ring and improved ventilation with openings in the hull roof had been prepared by 16 March 1944.

By 23 March, Prof. Dr. Porsche had asked Krupp for delivery of a second Maus I turret and the first Maus II turret. In May 1944, Krupp was given a contract to complete a 1/5th scale wooden model of a Maus II turret with the 7.5 cm Kw.K. mounted above the 12.8 cm Kw.K. At this time, Krupp was also redesigning the breech of the 7.5 cm Kw.K. for horizontal closure and designing a spent cartridge fume extractor fan for the 12.8 cm ammunition stowage in the Maus II turret.

Starting in the Fall of 1942, attempts were made to mount the 7.5 cm Kw.K. L/70 gun created for the Panther into a superstructure on a Sturmgeschuetz. On 12 September 1942, development work began on a "Sturmgeschuetz auf Leopard (L/70)". But, this effort was short-lived when the "Leopard" project was cancelled. (Refer to Panzer Tracts 20-2)

In early December 1942, Hitler ordered that work be accelerated on upgrading the Sturmgeschuetz from a 7.5 cm L/48 to a 7.5 cm L/70. As seen in the photos, Alkett created a full-scale wooden model to mount a 7.5 cm Pak L/70 on a Sturmgeschuetz chassis. The superstructure would have had to be completely redesigned to house the longer gun. (APG)

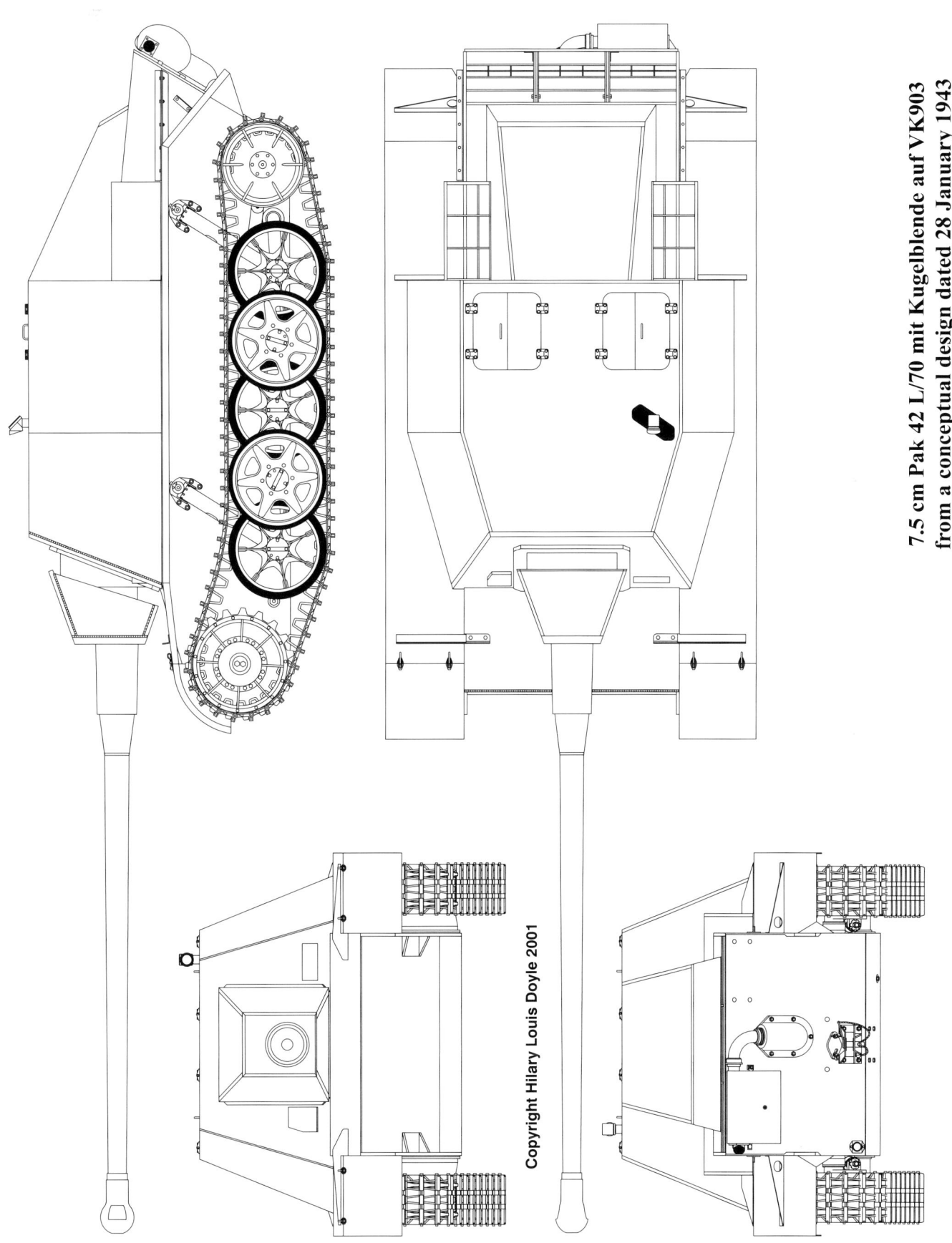

7.5 cm Pak 42 L/70 mit Kugelblende auf VK903 from a conceptual design dated 28 January 1943

Copyright Hilary Louis Doyle 2001

Panzerkampfwagen IV mit 7.5 cm Stu.G.40

On 5 February 1943, Krupp reported to Wa Pruef 6 that they had investigated the Munitionsministerium proposal to convert the Pz.Kpfw. IV into a Sturmgeschuetz. Krupp stated that this was not an acceptable conversion, since no weight would be saved by eliminating the turret. Krupp had based this comparison on their new 9./B.W. chassis design with sloped frontal armor (later rejected because of concerns that conversion to a different chassis design would disrupt production). The 7.5 cm StuK 40 L/48 was mounted in a superstructure adopted from the Sturmgeschuetz Ausf.F with limited traverse of 20 degrees and an elevation arc from -6 to +20 degrees. Sighting, access, and ventilation arrangements remained unchanged from the Sturmgeschuetz.

While the roof and rear of the superstructure design were copied from the Sturmgeschuetz Ausf.F design, the sides were increased to 45 mm thick. Armor protection for the hull was increased by sloping the 50 mm thick glacis plate at 56 degrees and increasing the hull front to a single 80 mm thick plate still set at 12 degrees.

The automotive drive train was taken over from the Pz.Kpfw.IV with a new final drive design from the 9./B.W. The 560 mm wide tracks resulted in a ground pressure of 0.76 kilogram/centimeter squared for the 28.26 metric ton Sturmgeschuetz. Overall length was 6.250 meters, width 3.260 meters, height 2.180 meters; and ground clearance 0.400 meter.

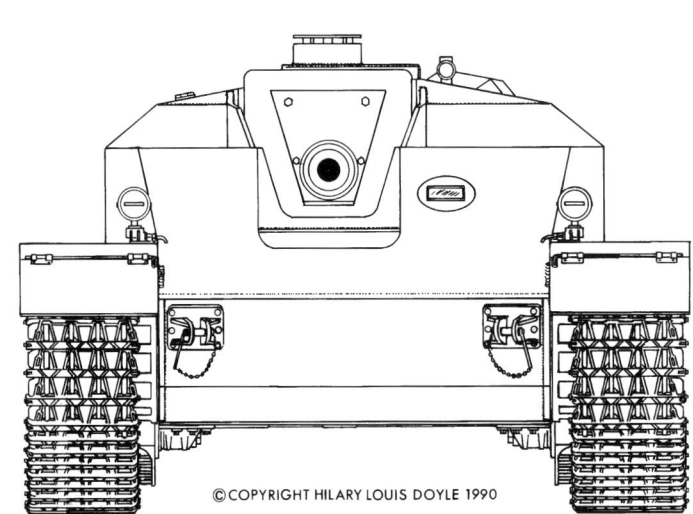

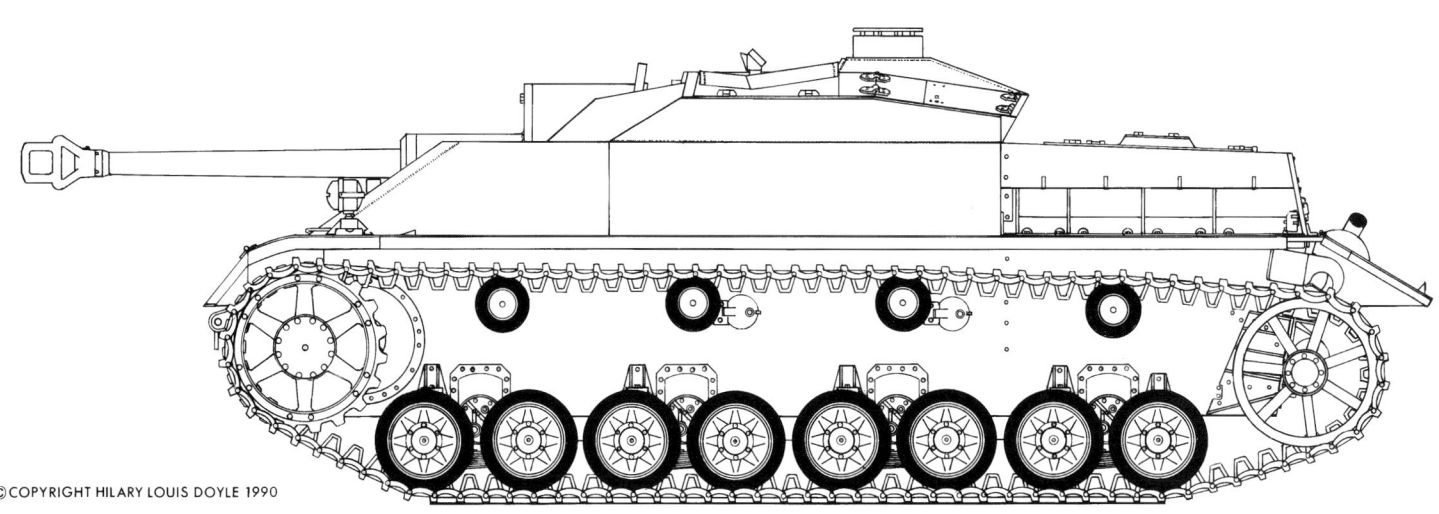

30.5 cm L/16 auf Sfl. Bär

On 4 March 1943, Krupp proposed a conceptual design for a "Sturmgeschuetz" mounting a 30.5 cm L/16 mortar and completed a conceptual design drawing known simply as the "Bär" (Bear) on 10 May 1943. The 30.5 cm Geschuetz was mounted in the superstructure with an elevation arc of 0 to 70 degrees and limited traverse of 4 degrees (2 degrees right or left). The gun with carriage and gun mantle weighed 16.5 tons. Two types of shells were to be fired. The Sprenggranate (high-explosive shell) weighed 350 kg, requiring 50 kg of propellant to achieve a muzzle velocity of 355 m/s for a maximum range of 10.5 kilometers. The Betongranate (concrete-piercing shell) weighed 380 kg, requiring 35 kg of propellant to achieve a muzzle velocity of 345 m/s for a maximum range of 10 kilometers. Recoil energy equaled 160 tons with the gun recoiling a distance of 1 meter. Only about 10 rounds were carried. This "Sturmgeschuetz" was to have a crew of six: commander, gunner, two loaders, driver, and radio operator.

The "Bär" was very well protected against most enemy anti-tank guns, with 130 mm thick armor on the upper hull front, 100 mm on the lower hull front, and 80 mm on the sides. The forward belly plate was to be 60 mm (rear 30 mm) to defeat the effects of anti-tank mines. 50 mm top armor was proof against high-explosive shells of up to 15 cm.

The chassis was designed by utilizing Tiger and Panther components (Maybach HL 230 engine rated at 700 metric horsepower at 3000 rpm, Zahnradfabrik AK 7-200 transmission, and Henschel dual radius L 801 steering unit) to achieve a maximum speed of about 20 km/hr. The suspension was designed by Krupp with ten 800 mm diameter rubber-cushioned steel-tired roadwheels on each side sprung with leaf springs. Two different tracks were envisaged - 500 mm wide transport track and 1000 mm wide off-road tracks. The 120 metric ton weight was dispersed along a track contact length of 5900 (when the tracks sank in 20 cm), resulting in ground pressure of 1.02 kilograms per square centimeter. A fairly poor steering ratio of 1.71 was achieved with the wheel centers 3100 mm apart. The overall length was 8200 mm, width 4100 mm, height 3550 mm, and ground clearance 500 mm.

On 10 May 1943, Krupp mentioned that Alkett was working on a 38 cm Sturmgeschuetz in competition with their "Bär". Alkett's design was selected for production as the Sturmmoerser on the Tiger I chassis (refer to page 8-56 of Panzer Tracts 8).

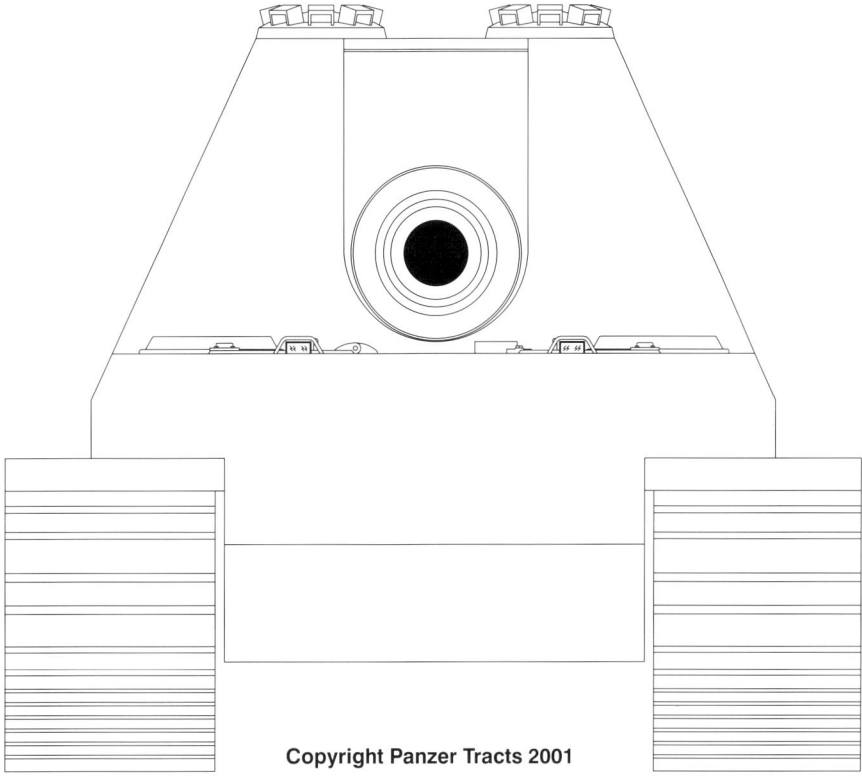

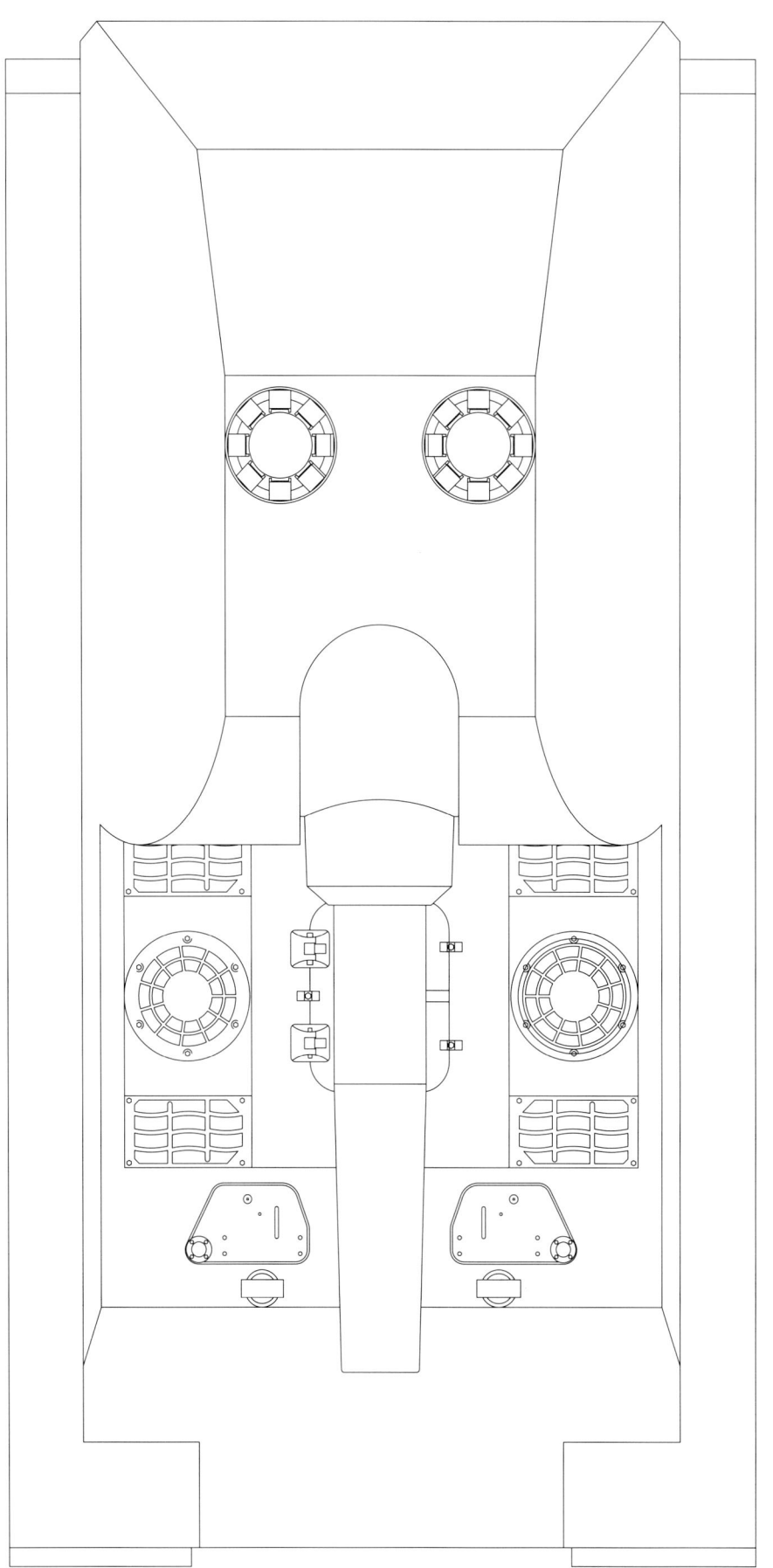

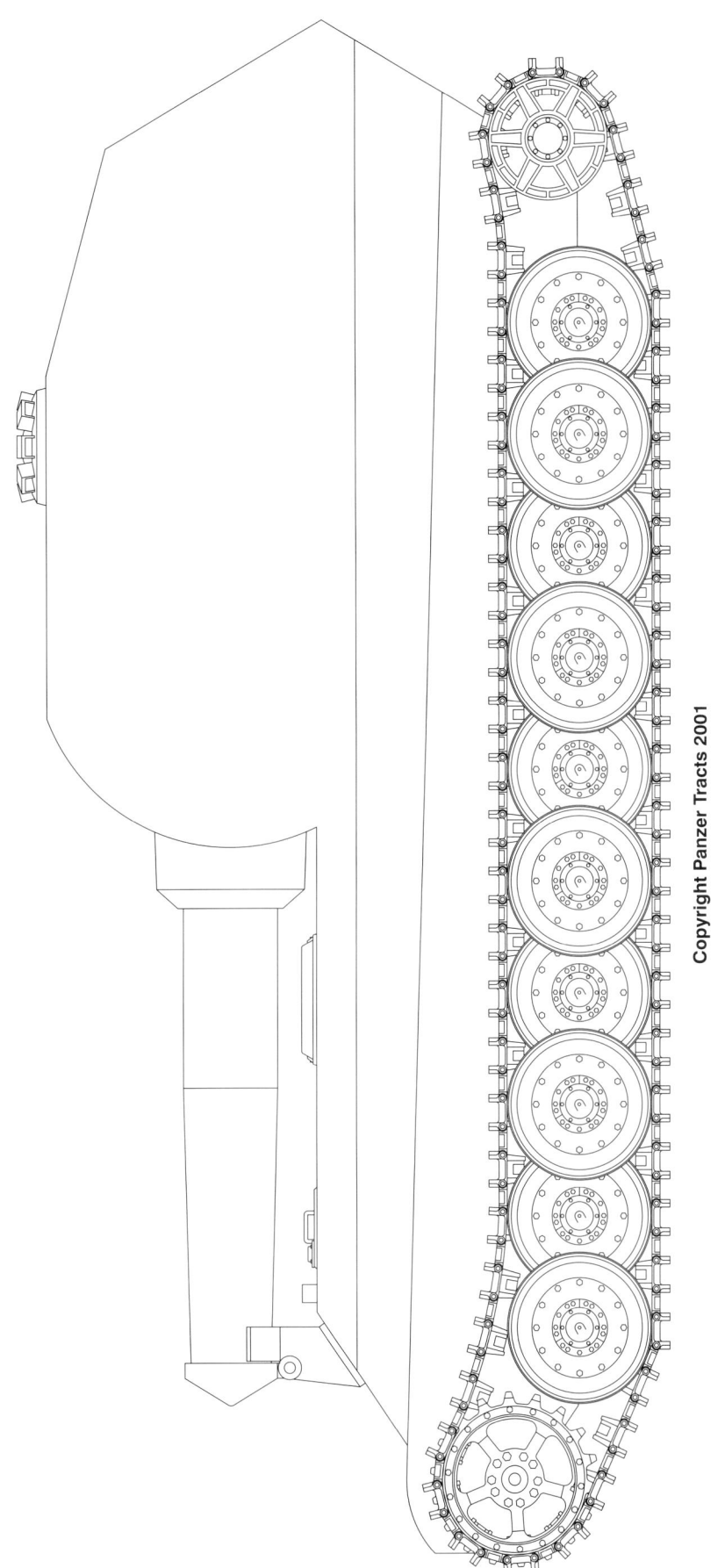

Krupp's conceptual design for the 30.5 cm L/16 auf Sfl. "Baer" (Bear)

Schwerer kleiner Panzerkampfwagen mit 10.5 cm

Porsche embarked on a totally new design, which they called a Panzerkampfwagen even though it was actually a Sturmgeschuetz with cast armor. Their first design of the "Panzerkampfwagen Porsche" was discussed with Rheinmetall who were designing the armament on 13 May 1943. The conceptual drawing was of a short compact vehicle with the 10.5 cm Sturmgeschuetz mounted in the hull front at a firing height of 1.37 meters. Sixty rounds of fixed hollow charge ammunition were carried to be fired at a maximum rate of 10 rounds per minute. Traverse was limited to 8 degrees to the right and left with an elevation arc of -7 to +15 degrees. Secondary armament hadn't been specified, but they wanted loop holes for firing machineguns and small arms for close defense to the sides and rear. It was to be manned by a crew of four - driver, gunner, loader, and a commander. The commander had an all-round vision cupola with ten periscopes. An escape hatch was to be located in the belly.

The cast armor was to be fairly thick, with 120 mm in the front, 80 mm for the sides, rear, and top, 50 mm belly up to the commander's seat, and 30 mm belly aft. Total weight was estimated to be 25 metric tons, of which 5.5 tons would be taken up by the drive train and suspension with tracks.

It had a low silhouette with average height of 1.95 meters but a fairly high ground clearance of 50 cm. It was considered to be an extraordinarily maneuverable Panzer due to a steering ratio of only 1.04 (short track contact length of 2.75 meters with a wheelbase of 2.65 meters).

During a meeting of the Panzerkommission on 15 May 1943, it was brought up that Hitler had agreed with Generaloberst Guderian's suggestion that every Panzer needed anti-aircraft defense. Rheinmetall proposed their MK 108 developed for the Luftwaffe which because of its small size would be especially suitable for this purpose. The Waffenamt stated that this weapon was affected by heat and dust.

This anti-aircraft weapon had been mounted in the commander's cupola in Rheinmetall's conceptual design drawing of the "schwerer kleiner Panzerkampfwagen mit 10.5 cm le.F.H.43 und 3cm automatische Fliegerabwehrkanone MK 108, dated 29 July 1943. The 3 cm MK 103 fired 0.33 kg high-explosive shells at a muzzle velocity of 525 m/s. Two versions of the cupola were shown, a fixed cupola with the MK 108 pointed to the rear, and a rotating cupola traversable through 360 degrees with an elevation arc from -5 to +90 degrees. 700 rounds of 3 cm ammunition were carried in seven steel link belts. Ammunition stowage for the 10.5 cm had been reduced to 44 rounds. Even though side, rear and top armor had been reduced from 80 to 70 mm, the overall weight was now estimated to be 27 metric tons. Six 780 mm diameter roadwheels rode on 600 mm wide tracks, resulting in a ground pressure of about 0.8 kg/cm².

The cast armor hull had been redesigned on the conceptual drawing of the "schwerer kleiner Panzerkampfwagen mit 10.5 cm le.F.H.43 und 3cm automatische Fliegerabwehrkanone MK 108" dated 3 February 1944. The thickness of the cast armor had been changed to 80 mm for the upper hull front, 60 mm for the lower hull front, roof, and rear, 40 mm for the slant sides, and 40 mm with 20 mm skirts on the vertical lower sides. Outer dimensions had also changed to 6.75 meters long, 3.15 meters wide, 2.20 meters high, and 0.45 meter ground clearance. With side panniers, ammunition stowage for the 10.5 cm le.F.H.43 had been increased to 77 rounds at the expense of 3 cm ammunition down to 600 rounds. Six 600 meter diameter roadwheels were mounted in pairs on each side. With 550 mm wide tracks at a ground contact length of 2.90 meters, ground pressure for the 26.6 metric ton Panzer had remained at about 0.8 kg/cm². The drive train was designed by Porsche with a rear sprocket drive powered by an air-cooled, 15.06 liter Porsche V-10 gasoline engine rated at 345 horsepower.

In the conceptual design dated 7 April 1944, the chassis remained unchanged but the main armament had been changed to a 10 cm PAW firing 10.5 cm 6.15 kg fin-stabilized hollow charge rounds at 900 m/s. Traverse was increased slightly to 10 degrees left and right of center. Only 56 rounds of ammunition were carried for the main gun.

The last changes found to this conceptual design were dated 1 June 1944. An infrared main gunsight had been added and a 1.28 meter base stereoscopic range finder mounted in the cupola.

The initial design for a schwerer kleiner Panzekampfwagen mit 10.5 cm from May 1943

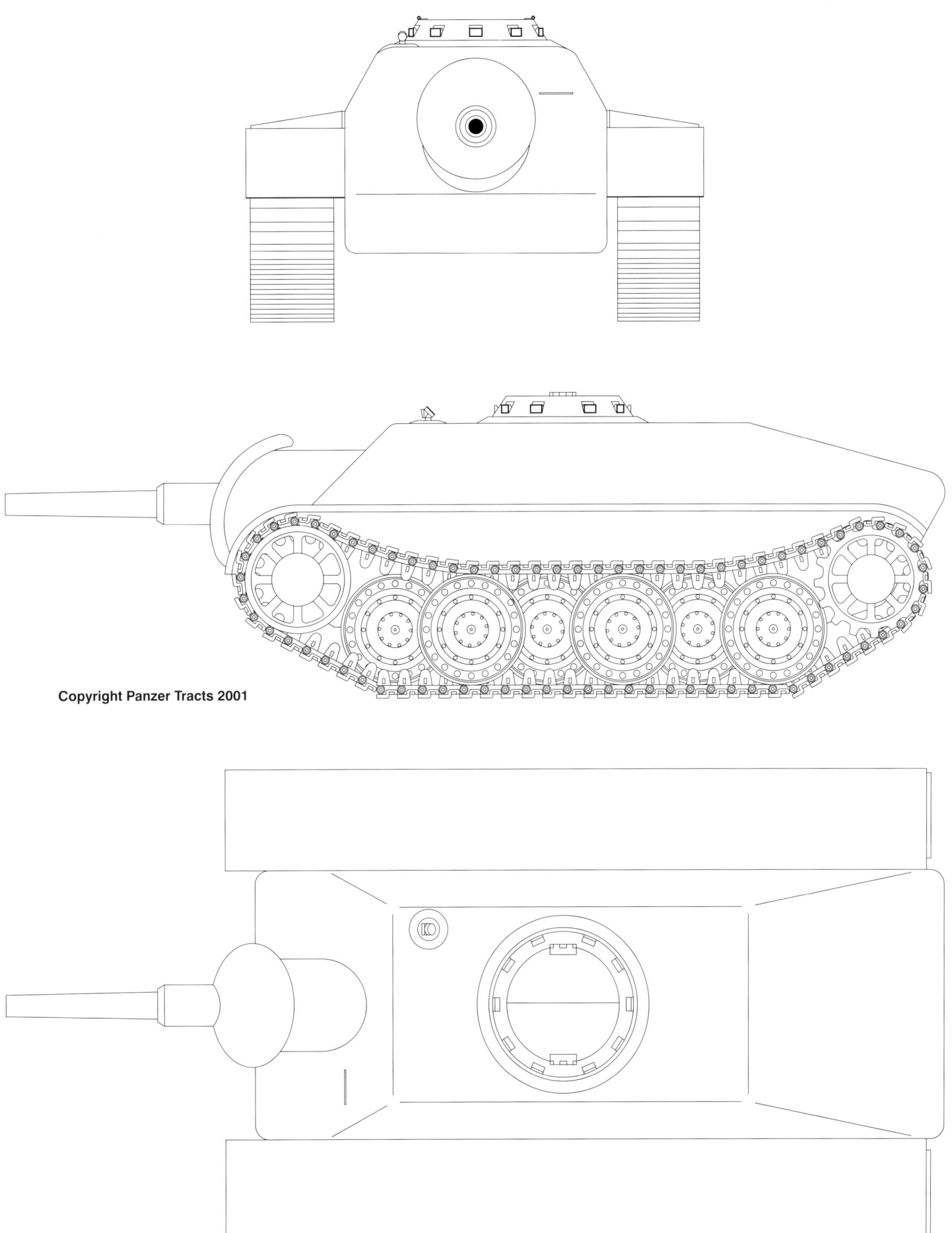

The schwerer kleiner Panzerkampfwagen mit 10.5 cm le.F.H.43 und 3 cm MK 108 from February 1944.

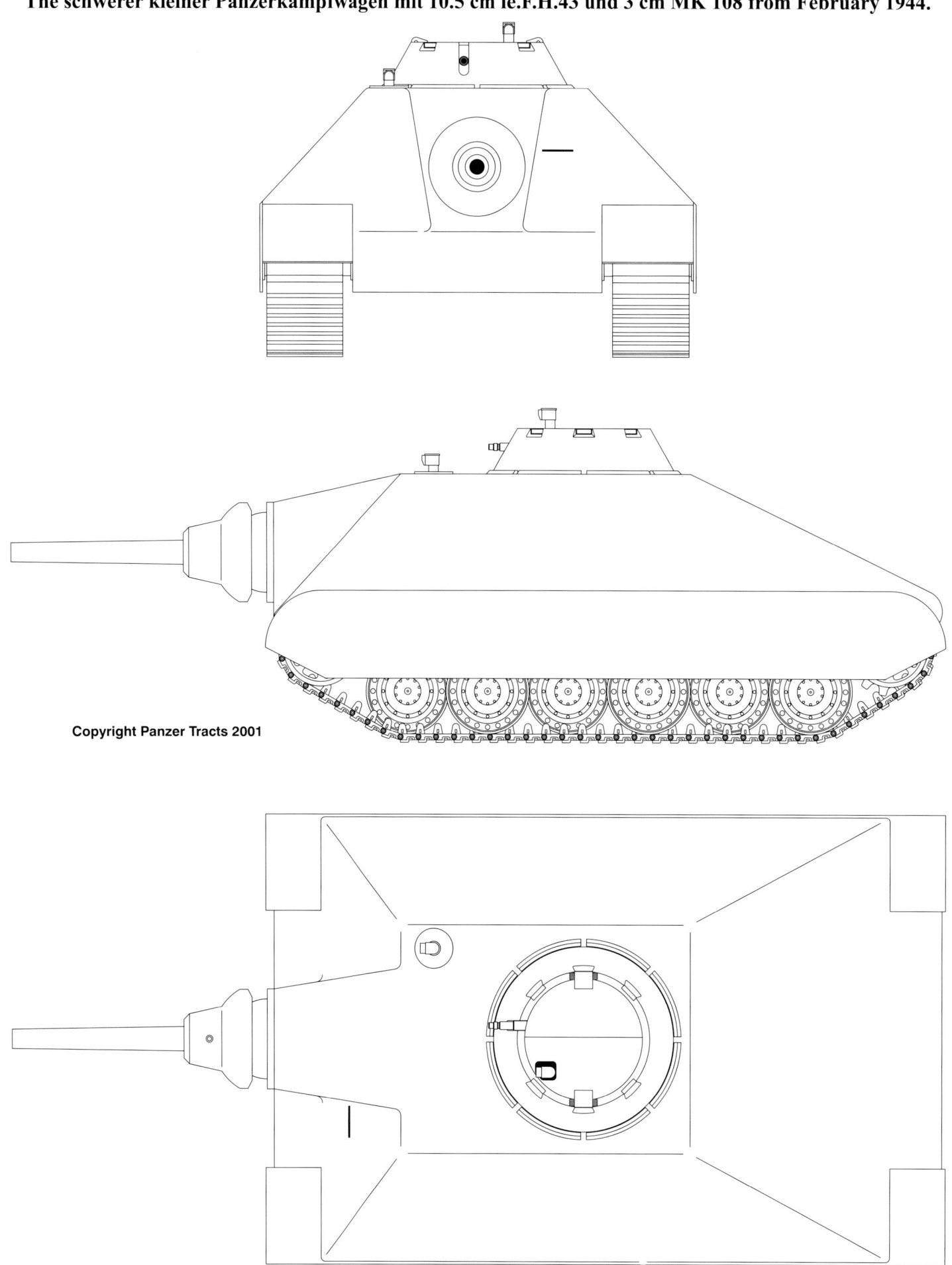

The main armament had changed to a 10 cm PAW in the drawing dated April 1944

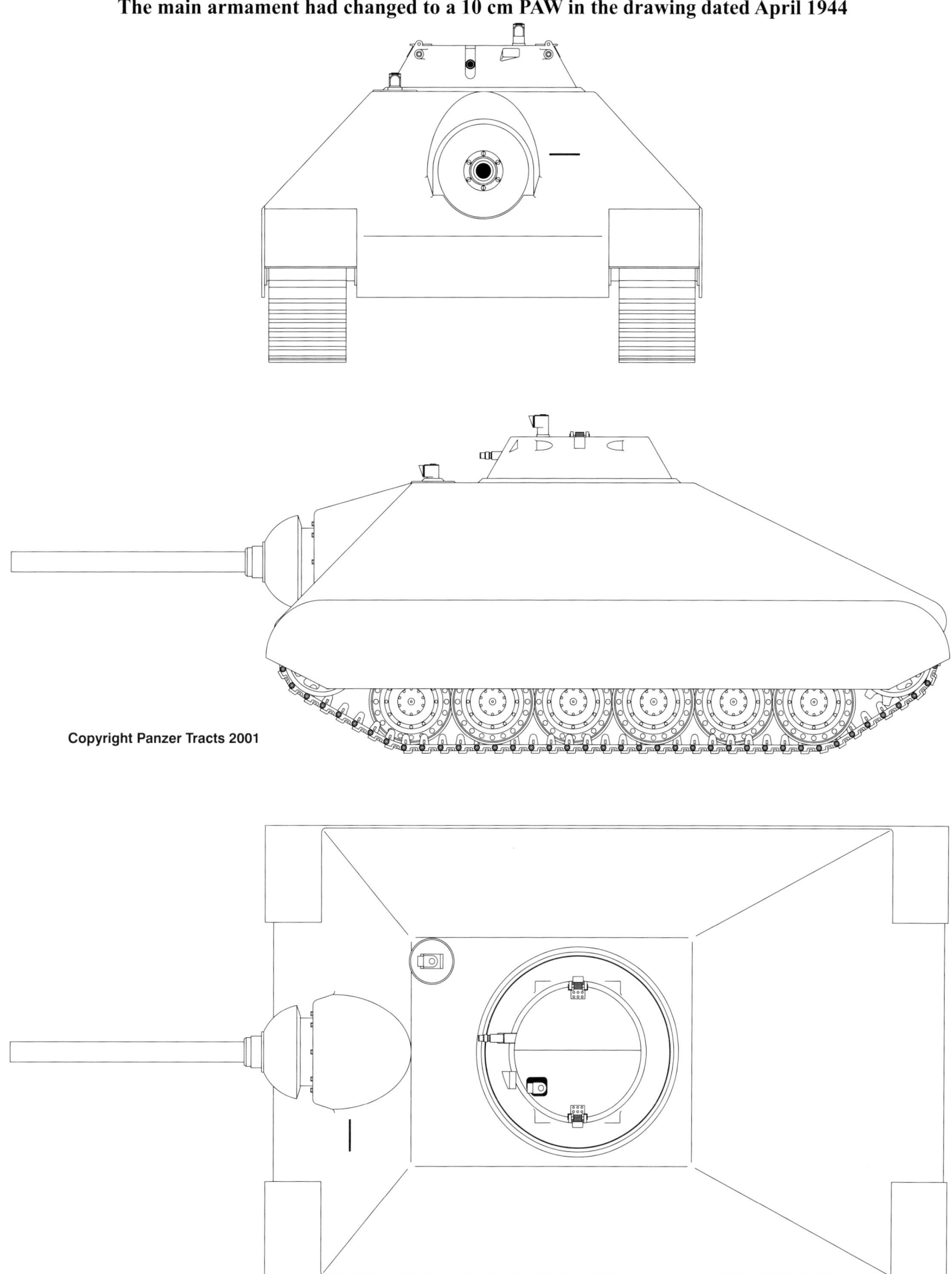

Rutscher
Panzerkleinzerstoerer

In late 1943, Wa Pruef 6 awarded contracts to the firms of BMW and Weserhuette for detailed designs of a klein Panzerjaeger (small tank destroyer) known as the "Rutscher". However, these projects were dropped in late February 1944 because the leichte Panzer-Jaeger 38 (refer to Panzer Tracts 9) was to fulfill their role as a cheap, mass-produced tank destroyer.

Nevertheless, the Entwicklungskommission Panzer (tank development committee) reactivated these projects when they met on 23 January 1945. General Major Thomale (Generalinspekteur der Panzertruppen) based the specifications for development of a Panzerkleinzerstoerer (small tank destroyer) on a fast, maneuverable anti-tank weapon to be used by the infantry as a close-combat weapons carrier. In their present situation the Panzerkleinzerstoerer was viewed as having special importance as being the weapon that could inflict the most damage on the enemy while expending the least amount of scarce raw materials and labor.

The specifications included a two-man crew, limited armor (20 mm front and 14.5 mm sides), low overall height of about 1.50 meters, high ground clearance of about 35 centimeters, high speed with a powerful engine of about 90 horsepower, low weight of 3.5 to 5 metric tons, and armed with a PAW and one machinegun. Five trial vehicles were to be rapidly completed.

Oberst Holzhaeuer (Wa Pruef 6) stated that about one year previously Wa Pruef 6 had worked through about 20 different proposals for a fast, small armored vehicle. Weserhuette had completed a design weighing about 3.5 tons using components especially designed for it, and Buessing had completed a 5 ton vehicle using available components.

Dr. von Heydekampf stated that development with available components could only result in a compromise solution, while an ideal solution could be achieved with specially designed components. The ideal solution must be strived for in order to defeat the American tank destroyer "Hellcat" with its torsion bar suspension and top speed of 90 km/hr. The Entwicklungskommission decided to forcibly pursue both development paths.

As reported on 23 January 1945, the PWH 8 H 63 was developed at the request of the General-inspekteur der Panzertruppen. It was not a recoilless gun, but had a smooth bore for firing fin-stabilized projectiles at 520 meters per second. The shaped charge anti-tank round could penetrate 140 to 150 mm of armor at 30 degrees. Its effective range was about 700 meters, with a maximum range of 1500 meters possible.

The firm of Zahnradfabrik Friedrichshafen was working on developing two different transmissions (a 5-speed FAK-45 transmission and a semi-automatic, easily shiftable transmission) to be coupled to a 150 horsepower Saurer diesel engine in a Panzerkleinzerstoerer in March 1945.

On 19 March 1945, the Generalinspekteur der Panzertruppen recorded the following comments on development of a Panzerkleinzerstoerer: A vehicle with the required characteristics, especially a total weight of 3.5 tons, requires completely new components (engine, drive train, suspension, etc.) whose development and series production program can be achieved at the earliest in 1-1/2 to 2 years. Therefore, due to time constraints, it shouldn't be pursued further. Only a vehicle weighing about 7 to 10 tons can be quickly developed and quickly brought into mass production using available components currently in mass production.

An especially small (maximum 5 tons), fast and maneuverable vehicle is needed because the PWK 8 H 63 has an effective range of only 600 to 700 meters. Under the current circumstances, a 7 to 9 ton vehicle (that must close to within 700 meters to engage heavily armored and armed opponents' tanks) is rejected by the Generalinspekteur der Panzertruppen as tactically worthless.

Therefore, in view of the current development and production situation, the "Panzerkleinzerstoerer" project must be dropped. The contracted trial vehicles (3.5 ton Weserhuette and 7.5 ton Daimler-Benz) are to be completed and delivered for automotive and tactical testing. In addition to the PWK weapon, an attempt should be made to install the 7.5 cm Kw.K. L/48 to engage targets on the battlefield at ranges over 700 meters.

The wooden model of a "Rutscher" as conceived by BMW (WJS)

"Hetzer"
Entwicklungsfahrzeug E 10

The E-Serie program was conceived by Oberbaurat Kniepkamp (civilian head of automotive design in Wa Pruef 6) in May 1942 and authorized as a project in April 1943. Contracts for the smallest Panzer in the series, the E 10, were awarded to the Klockner-Humboldt-Deutz Magirus Werk in Ulm.

The E 10 was primarily developed as a Panzer chassis to test new components, especially engines, transmissions, and suspensions. Many parts were to be shared by the E 10 and the E 25. According to Oberst Holzhaeuer (head of Wa Pruef 6) the reason for converting to rear drive was to compensate for heavy frontal armor.

A Magirus drawing shows the E 10 outfitted with the 7.5 cm Pak 39 L/48. The type of weapon for series production was still unsettled in January 1945 but was to be based on the most modern weapon available.

Armor protection was to consist of a 60 mm thick glacis plate sloped at 60 degrees from vertical, 30 mm lower hull front at 60 degrees, 20 mm sides at 10 degrees, 20 mm upper hull rear at 15 degrees, 20 mm lower hull rear at 33 degrees, and 10 mm deck and belly plates. The hull rear was hinged and could be unbolted to remove and replace the entire drive train including engine within a few hours.

The Maybach V-12 HL 100 engine, rated at 400 metric horsepower, was to be installed in the Versuchsfahrgestell. Maybach was adding fuel injectors and improving cooling to create the HL 101 rated at 550 horsepower at 3800 rpm for series production. With this powerful engine, the E 10 was to be especially suitable for employment as a "Hetzer" against the new fast enemy "Hellcat" tank.

Four large 1000 mm diameter rubber-cushioned steel-tired roadwheels were overlapped to reduce the track ground contact length to 2.55 meters which, combined with the wheelbase of 2.46 meters, provided an optimum steering ratio of 1.04. The E 10 was unique in having an auxiliary drive coupled to the suspension in order to lower and raise the vehicle. Ground pressure was exceptionally low due to 400 mm wide tracks.

Overall length with the 7.5 cm Pak 39 L/48 gun was 6.91 meters, chassis length 5.35 meters, overall width 2.86 meters, and overall height to top of air louvers 1.76 meters.

According to Kniepkamp in postwar interrogations, the drawings for the E 10 were completed in the Summer of 1944 and contracts given to Magirus to complete three trial vehicles. The three hulls were being made in Silesia, but were not finished when the Russians arrived.

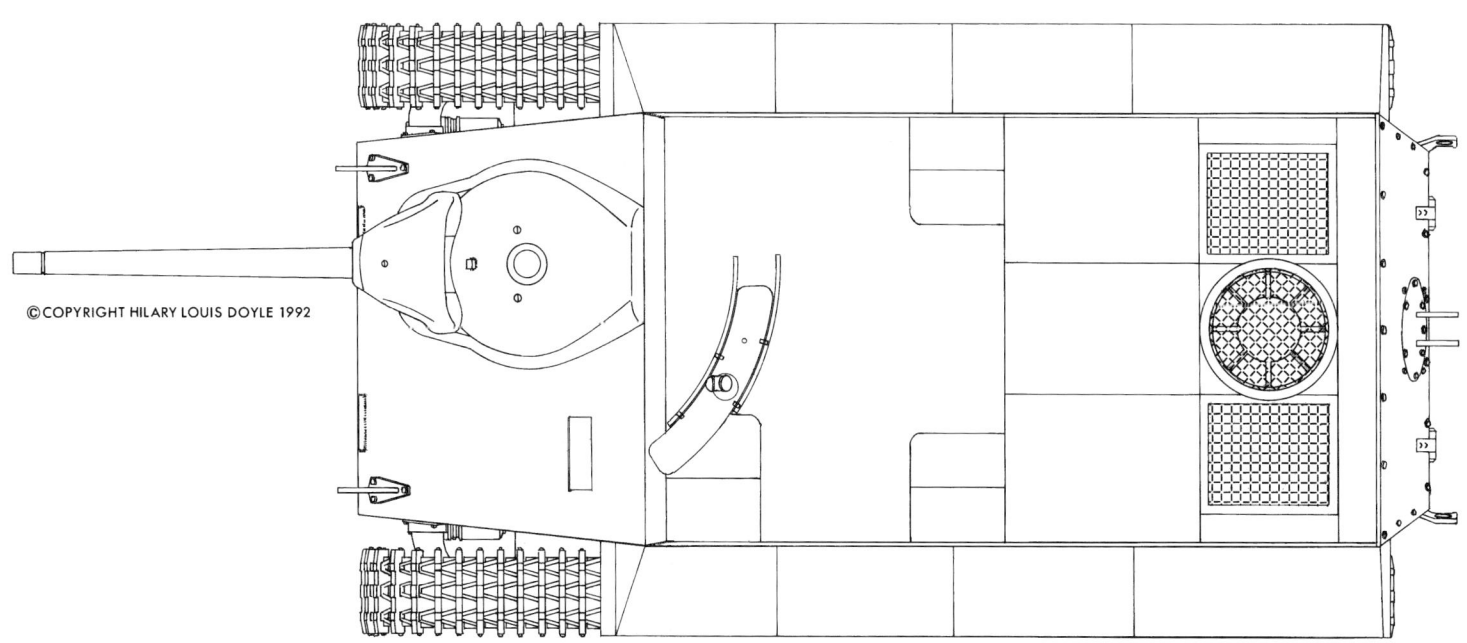

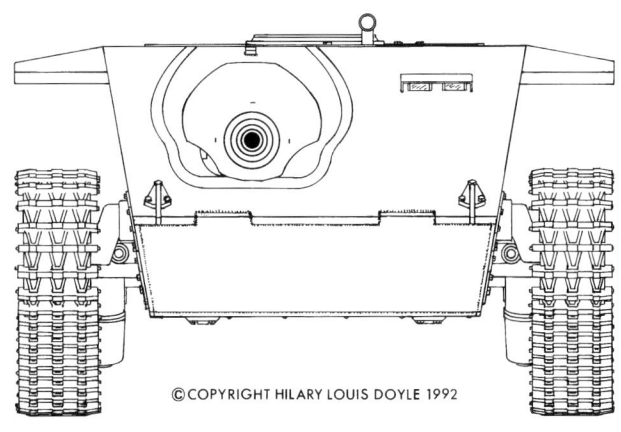

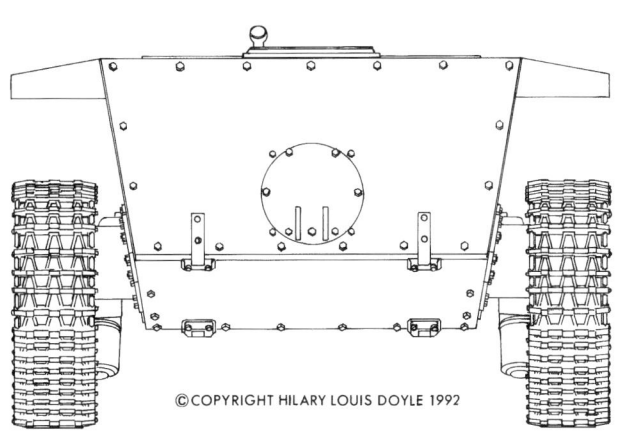

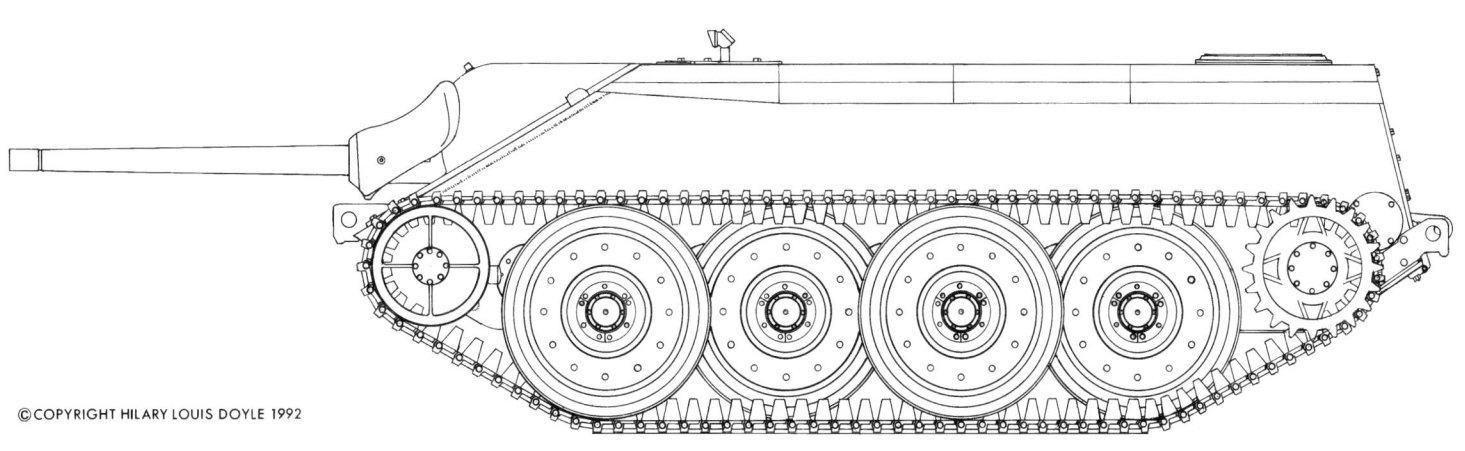

20-41

Entwicklungsfahrzeug E 25

The E 25 was the second largest Panzer in the Entwicklungsfahrzeug series created by Oberbaurat Kniepkamp of Wa Pruef 6. The detailed design contract was awarded to Argus Werke in Karlsruhe, with the design project directed by Dr. Klaue. The E 25 was to have rear drive and share many parts with the E 10. In January 1945, the Entwicklungskommission decided that this design was still worth pursuing as it again filled a vacancy in the Panzerprogramm for vehicles in the 25 ton class.

A drawing shows the E 25 outfitted with a 7.5 cm Pak L/70 gun mounted on the glacis plate. Selection of a weapon for series production hadn't been determined in January 1945.

Armor protection consisted of the 50 mm thick glacis plate sloped at 50 degrees from vertical, 50 mm lower hull at 55 degrees, 30 mm upper hull side at 52 degrees, 30 mm lower hull side at 0 degrees, 30 mm upper hull rear at 40 degrees, 30 mm lower hull rear at 50 degrees, and 20 mm roof and belly.

On 1 July 1944, Zahnradfabrik-Friedrichshafen was working on the design of a semi-automatic transmission/steering/final drive block for the E 25 to be completed in early 1945. This drive train block was to be coupled to the Maybach V-12 HL 100 engine. The more powerful Maybach HL101 with fuel injection was listed as the current choice in late March 1945.

Hull length was 5.66 meters, width 3.41 meters, height 2.03 meters, and ground clearance 0.51 meter. Tracks were about 700 mm wide. The ground contact length of 2.96 meters and wheelbase of 2.74 meters resulted in a steering ratio of 1.08.

Only a few Versuchsfahrgestell were ordered. The design had not proceeded far enough to consider mass production. According to Oberbaurat Kniepkamp in postwar interrogations, three E 25 hulls were reported to be at Alkett in Berlin-Spandau, but they were no longer there when Western Allies went to investigate.

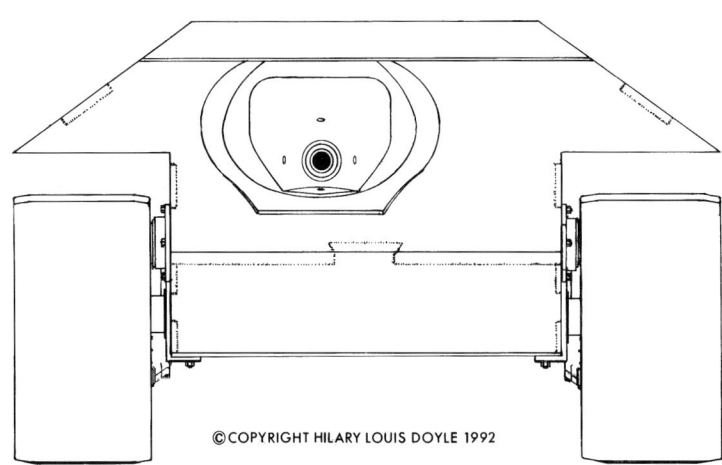

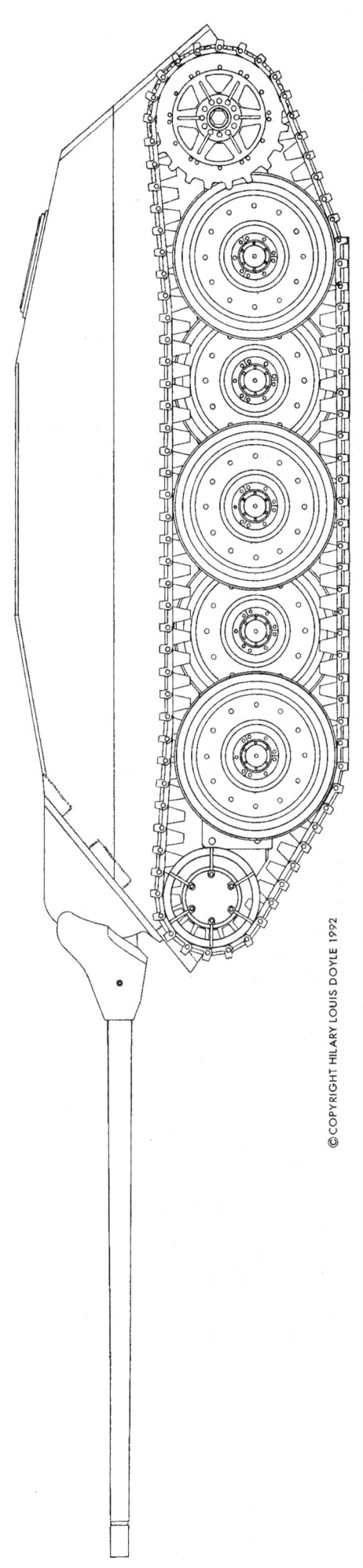

20-43

Panzer IV lang (E)

The Panzerkommission met on 4 January 1944 to decide on the specifications for a combination Pz.Kpfw. III/IV in order to meet the requirement from Hauptdienstleiter Saur of the Speer Ministry that medium and heavy Panzers be standardized. This same design was also to be used as the chassis for a leichte Panzer Jaeger III/IV. It had been determined by September 1943 that the 7.5 cm Pak 42 L/70 could be mounted in the superstructure of the leichte Panzerjaeger IV, and this same superstructure was adapted for the new Einheitsfahrgestell (standardized chassis).

Front hull armor was redesigned to come to a point, with the 60 mm thick glacis plate sloped at 60 degrees and 60 mm lower hull plate at 45 degrees. The 80 mm thick superstructure front plate was also well sloped at 50 degrees and the 30 mm thick side of the superstructure at 36 degrees. Schuerzen (steel armor skirts) were to be made out of 5 mm thick, flat interchangeable pieces.

The drive train was to retain the Maybach HL 120 TRM engine and cooling system from the Pz.Kpfw.IV in combination with the SSG 77 transmission, steering gear, and reinforced final drives and sprocket from the Pz.Kpfw.III. Various types of suspensions were tested before deciding on six 660 mm diameter rubber-saving steel-tired roadwheels, mounted in pairs with leaf springs. New symmetrical 540 mm wide track was selected with a center guide, consisting of paired cast and formed links (similar to Tiger II track). The idler wheel was retained from the Pz.Kpfw.IV, but the adjusting mechanism had a wider span and was strengthened and simplified. The driving range was significantly increased by adding a 300 liter fuel tank in the engine compartment.

Contracts were awarded in March 1944 for the production of three trial Einheitsfahrgestell by Alkett. By early May 1944, plans had been made to convert both Alkett and Miag production from the Sturmgeschuetz III to the Panzerjaeger III/IV starting in November 1944. Krupp was to convert from Sturmgeschuetz IV starting in January 1945, Vomag was to convert from Panzerjaeger IV starting in March 1945, and even Nibelungenwerk was to convert from the Pz.Kpfw.IV starting in March 1945. On orders from Hitler, the names of all the Sturmgeschuetz mounting 7.5 cm Pak L/70 guns were to be changed to Panzer IV lang. The Sturmgeschuetz auf Einheitsfahrgestell received the name Panzer IV lang (E).

Deutsche Edelstahl completed the armor body for a single leichte Panzerjaeger III/IV in September 1944, But the next month the Panzer III/IV program was dropped when on 4 October 1944 the Panzerkommission decided to reduce the number of different chassis for all armored vehicles to just three, the 38t, Panther, and Tiger, starting about the middle of 1945.

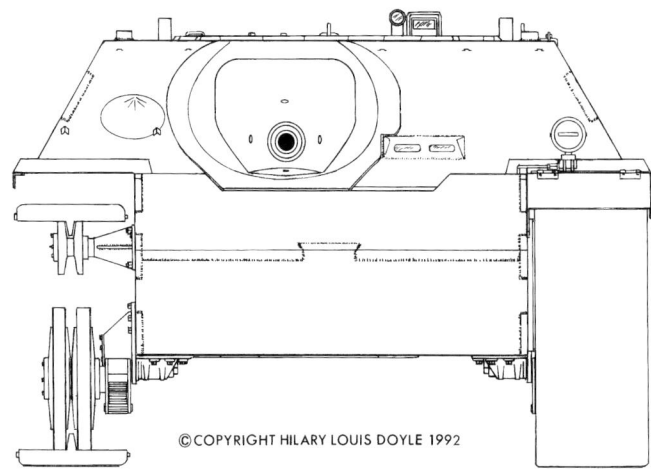

©COPYRIGHT HILARY LOUIS DOYLE 1992

©COPYRIGHT HILARY LOUIS DOYLE 1992

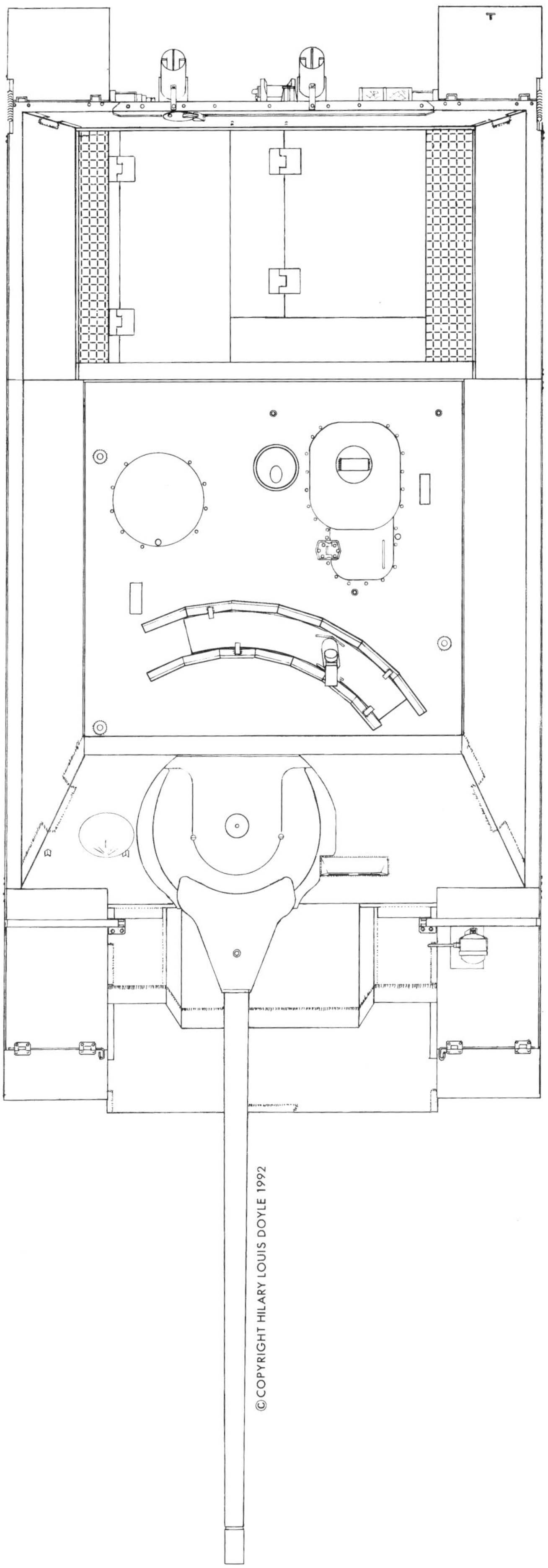

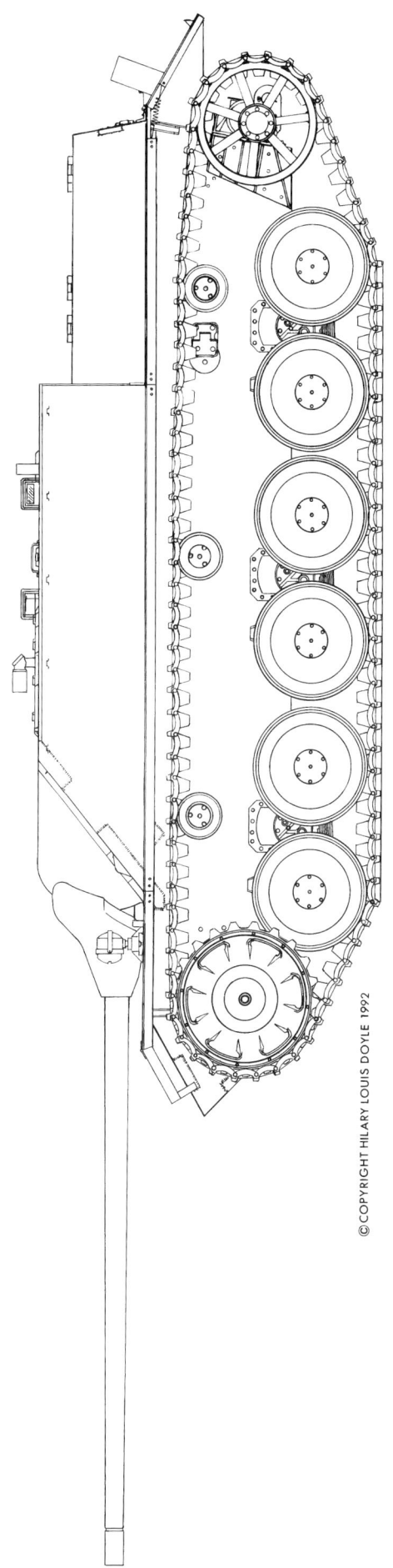

Jagdpanzer 38 D

On 26 September 1944, OKH announced that a new model of the leichte Panzerjaeger 38 t with a Tatra diesel engine was to be produced by Alkett at a maximum rate of 1000 per month. On 4 October 1944, the Panzerkommission decided to limit future production to only three types of Panzer chassis based on the 38t, Panther, and Tiger. Wa Pruef 6 revealed on 19 October 1944 that due to insufficient component deliveries, BMM and Skoda had fallen far short of production goals and this had created difficulties in supplying 38t chassis for other uses. The 38t chassis was poorly suited for production by German industry because of different machine tools such as those needed for stamping the inner gears for the final drives and stamping instead of grinding the beveled gears for the steering unit.

The Panzerjaeger was to be completely redesigned for the Reich. The new Ausfuehrung "Reich" was to have a new hull with vertical sides, a new suspension with stronger springs, a 60 mm wider wheel base, 220 horsepower Tatra diesel engine with transfer case, AK 5-80 transmission, a stronger steering unit, and stronger final drives. This detailed design project was assigned to Chef-Ing. Michaels of Alkett.

On 20 November 1944, the General der Artillerie under the Chef Gen.St.d.H. recorded that the new Sturmgeschuetz was to be named Jagdpanzer 38 "d". Design of the chassis was almost complete, and design of the superstructure and interior layout was to start shortly.

At a meeting of the Entwicklungskommission Panzer on 23 January 1945, Obering. Michaels explained that redesign of the 38t chassis was absolutely necessary to install the Tatra diesel engine and to create a mature design for mass production. The suspension and hull were significantly changed to allow future replacement of the current 350 mm wide tracks by 460 mm wide tracks without additional modification. A new suspension with vertical volute springs had been developed for the 38 D that was guaranteed to be suitable for 20 metric ton loads.

The drive train, with air-cooled V-12 Tatra diesel engine producing 200 horsepower at 2000 rpm, was designed for a maximum speed of 40 km/hr. A full tank of 380 liters of diesel fuel was equal to a range of 500 kilometers on roads or 300 kilometers cross-country. The steering ratio was 1.2 as compared to 1.27 in the 38t. With 62 rounds of ammunition, combat weight was about 16.7 tons. Three different weapon systems were to be mounted in the basic Jagdpanzer 38 D with 60 mm frontal armor: the Jagdpanzer 38 D mit 7.5 cm Pak 39 L/48; the Jagdpanzer 38 D mit 7.5 cm Pak 42 L/70 (about 500 to 600 kg heavier); and the Jagdpanzer 38 D mit 10 cm Sturmhaubitze 42/2.

Contracts were awarded to Alkett for completion of two Versuchsfahrzeug 38 D. As revealed on 30 January 1945, planned mass production of the Jagdpanzer 38 D was to start at Alkett with 5 completed in March and rapidly increase to 800 per month by December. Vomag was scheduled to complete their first 5 in July and increase to 300 per month by December.

On 14 March 1945, the Generalinspekteur der Panzertruppen revealed that, as planned, the first Jagdpanzer 38 D were to be completed with diesel engines in June 1945. This new model was to start production with a 7.5 cm Pak 42 L/70 with normal recoil, 50 rounds of ammunition, and 50 mm frontal armor. It was planned to demonstrate the first Versuchs-Jagdpanzer 38 D with a 7.5 cm Pak 42 L/70 starr for Hitler on 20 April 1945.

On 23 March 1945, Zahnradfabrik-Friedrichshafen reported that assembly of the first two 38 D Versuchsfahrzeuge at Alkett had advanced to the point where they could be running within 8 days after receipt of the transmissions. Because these vehicles had to be ready for delivery on 15 April, arrangements were made to send two AK 5-80 transmissions by truck on 5 April to arrive at Alkett for delivery to Obering. Michaels on 9 April 1945. It is not known whether these two 38 D were completed. In postwar interrogations, Alkett management stated that their records had been burned.

Plans had also been made to produce a reconnaissance version of the 38 D with a longer hull, thinner glacis armor, and a longer track contact length. This version known as the Aufklaerer 38 D will be covered in more detail in Panzer-Tracts 20-2.

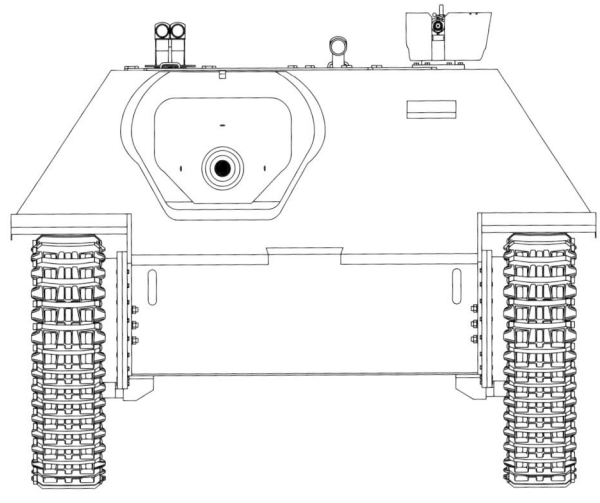

Copyright Panzer Tracts 2001

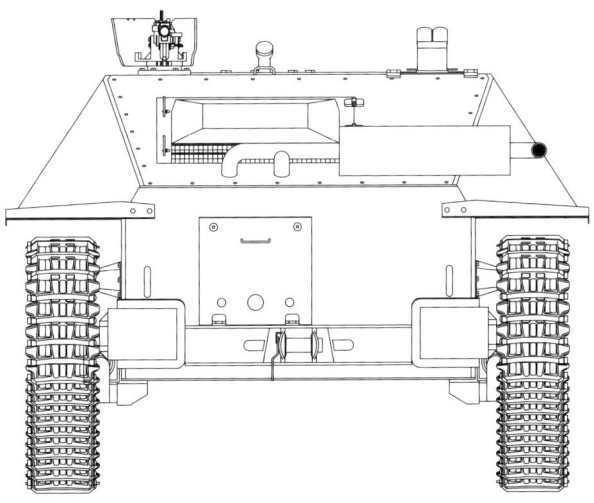

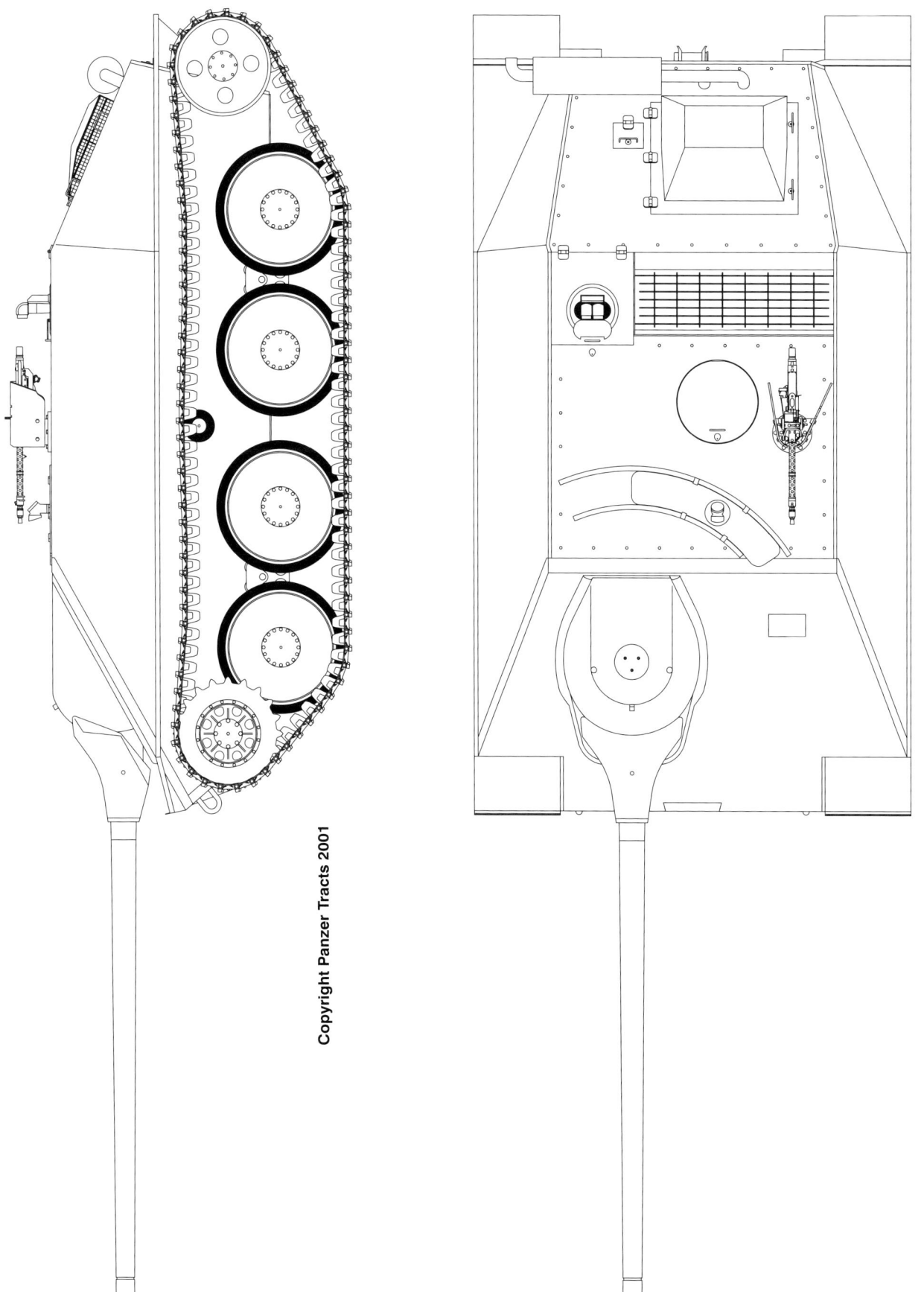

Alkett was working on the Jagdpanzer 38 D with a 7.5 cm Pak L/70 gun at the end of the war.

Krupp Proposals to "Umbewaffnung der Panzer"

In November 1944, Krupp prepared a series of conceptual designs proposing to increase the armament of all Panzers and Jagdpanzers currently in production. The sketches were sent to the Generalinspekteur der Panzertruppen, who in turn passed them to Wa Pruef 6. On 20 January 1945, Wa Pruef 6 attempted to instill some sense of reality with the following position paper on Krupp's proposal to "Umbewaffnung der Panzer" (rearm the tanks):

The Krupp proposals are based on engaging the enemy with the most effective weapon while using the least amount of armor. Krupp thinks that the question of increased weight from larger weapons can be solved by giving up the idea "that Panzers should have sufficient armor to protect them against the penetration capability of their own weapon." However, armor protection for Panzers is based mainly on the advancements that the opponent has achieved in his weapons technology and less on the capabilities of our weapons. Therefore, the heavier armament needed to obtain long-range penetration of thicker armor must be balanced with better or at least equivalent armor protection so that the Panzer with its expensive gun is not inferior to the opponents from the start.

These proposals mean - especially in the Jagdpanzer sector - a significant redesign of the models currently in production so that their realization can't be achieved without considerable disruption in production. Whether this can be justified under the current circumstances must be decided by higher positions.

Comments on the individual proposals were:

Pz.Kpfw.38 mit 7.5 cm KwK 40 (L/48) - It is possible to mount the Pz.IV turret within available dimensions, but it would require a steeper angle to the sides and front. Dropping the mechanical traversing gear would lower traversing speed so that there would be no advantage to this Panzer when compared to the current Jagdpanzer 38. A weight of 16 metric tons is an extreme disadvantage - the Jagdpanzer at 15 tons is already at the limit that can be supported by the roadwheels.

Jagdpanzer 38 mit 7,5 cm Pak 42 (L/70) - Because production of the Panzer IV lang is to cease, a Pak 42 mounted in a Jagdpanzer 38 is greatly desired; however, installation in the proposed design would create considerable difficulties. By relocating the fighting compartment and the engine, practically all parts would change, creating a new vehicle resembling the Pak 40 Sfl. Armor protection is totally insufficient. However, with sufficient armor it would weigh 18 tons, which can't be supported by the suspension. It would have about the same characteristics as the Panzer IV lang A, which is a failure based on reports from the front.

Pz.Kpfw.IV - Proposals based on the Pz.Kpfw.IV were not reviewed, since production of this chassis is to cease.

Pz.Kpfw.Panther mit 8.8 cm KwK 43 (L/71) - Installation of the 8.8 cm KwK43 in a Panther turret has already been attempted in a wooden model. The turret must have a 100 mm larger diameter than Krupp's proposal in order to have sufficient space for the loader. Instead of 103 rounds for the 7.5 cm KwK42, at most 45 to 50 8.8 cm rounds could be carried. The increased weight of about 1 ton is still supportable by the Panther chassis.

Jagdpanther mit 12,8 cm Pak 80 (L/55) - Fundamentally it is possible to install the Pak 80 in the Jagdpanther. However, an entirely new chassis would result due to relocating the fighting compartment. It would be a "lame duck" due to the extremely limited interior space and an estimated weight of 51 tons. In addition, the rate of fire would be considerably reduced with the two part 12.8 cm ammunition. In its current model, the Jagdpanther is a lively vehicle with a gun that will meet all requirements for a long time. Therefore, Krupp's proposal can't be approved.

Tiger B mit 10.5 cm KwK (L/68) - The proposed 10.5 cm gun was not accepted into service by the army. Therefore, it doesn't appear advisable to produce a special caliber just for this. In any case, the gun requires new traversing and elevation devices and probably the turret would have to be redesigned. The rate of fire would sink considerably because of two-piece ammunition. In addition, a second loader would be needed, and adding space for him would create considerable difficulty.

Jagdtiger mit 12,8 cm Pak (L/66) - Installation of a longer 12.8 cm gun in the Jagdpanzer was already investigated. The engine compartment location had to be rejected. Due to railway profile restrictions, the gun can't be mounted high enough to allow sufficient depression of the gun tube. Moving the fighting compartment forward results in a gun overhang which interferes with driving safety. By moving the gun back, as proposed by Krupp, the vehicle loses its characteristics as a Jagdpanzer which can quickly open fire, and can only be referred to as heavy armored artillery.

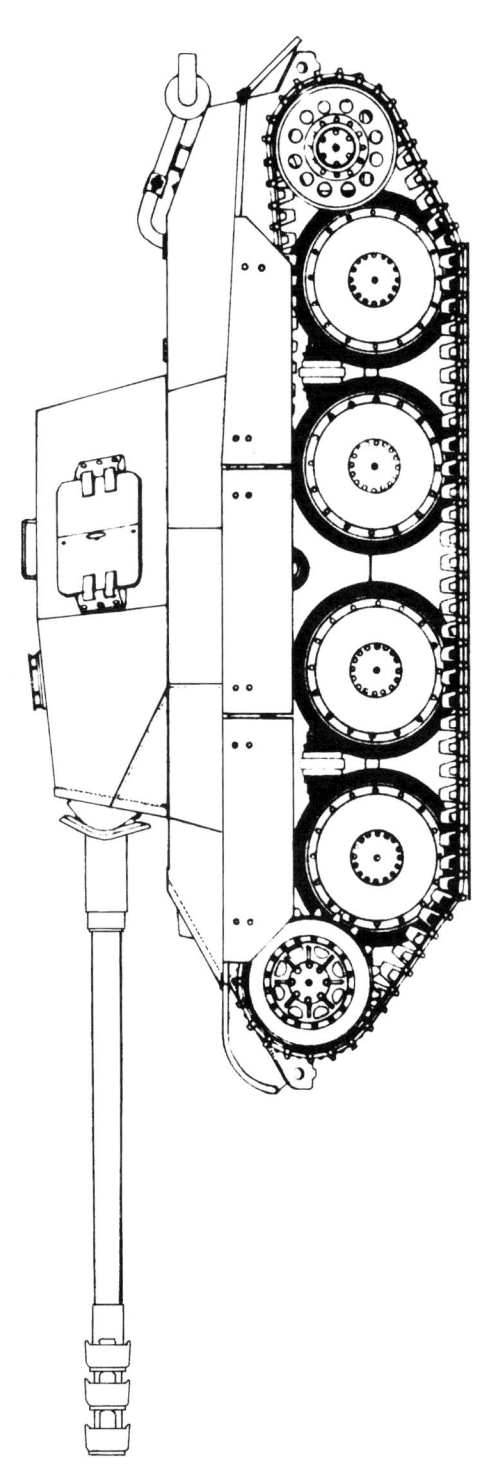
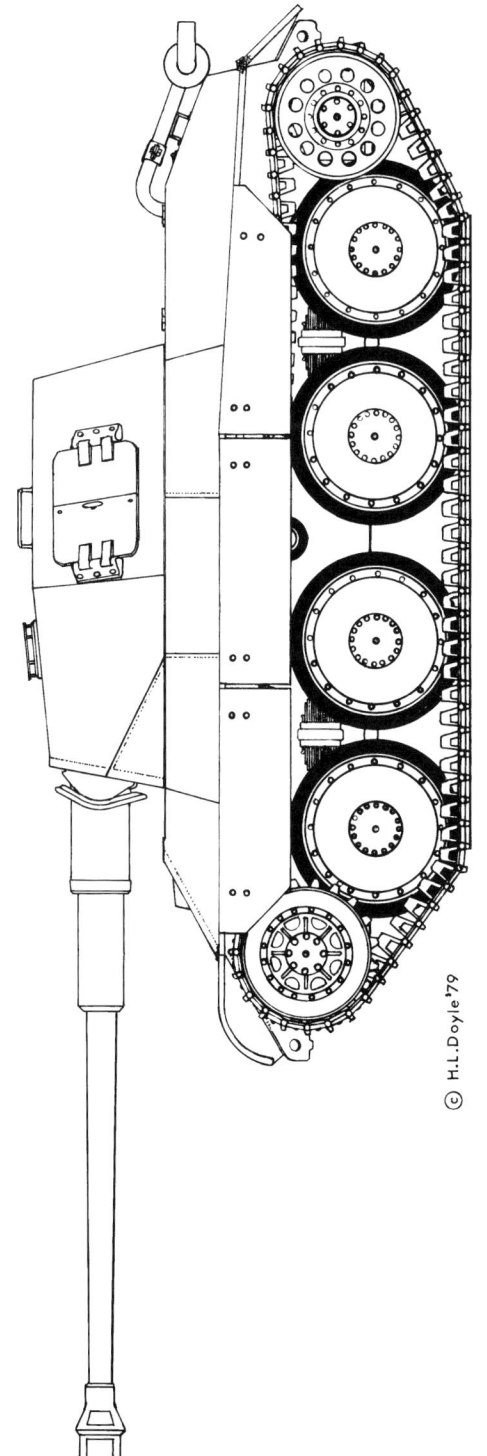

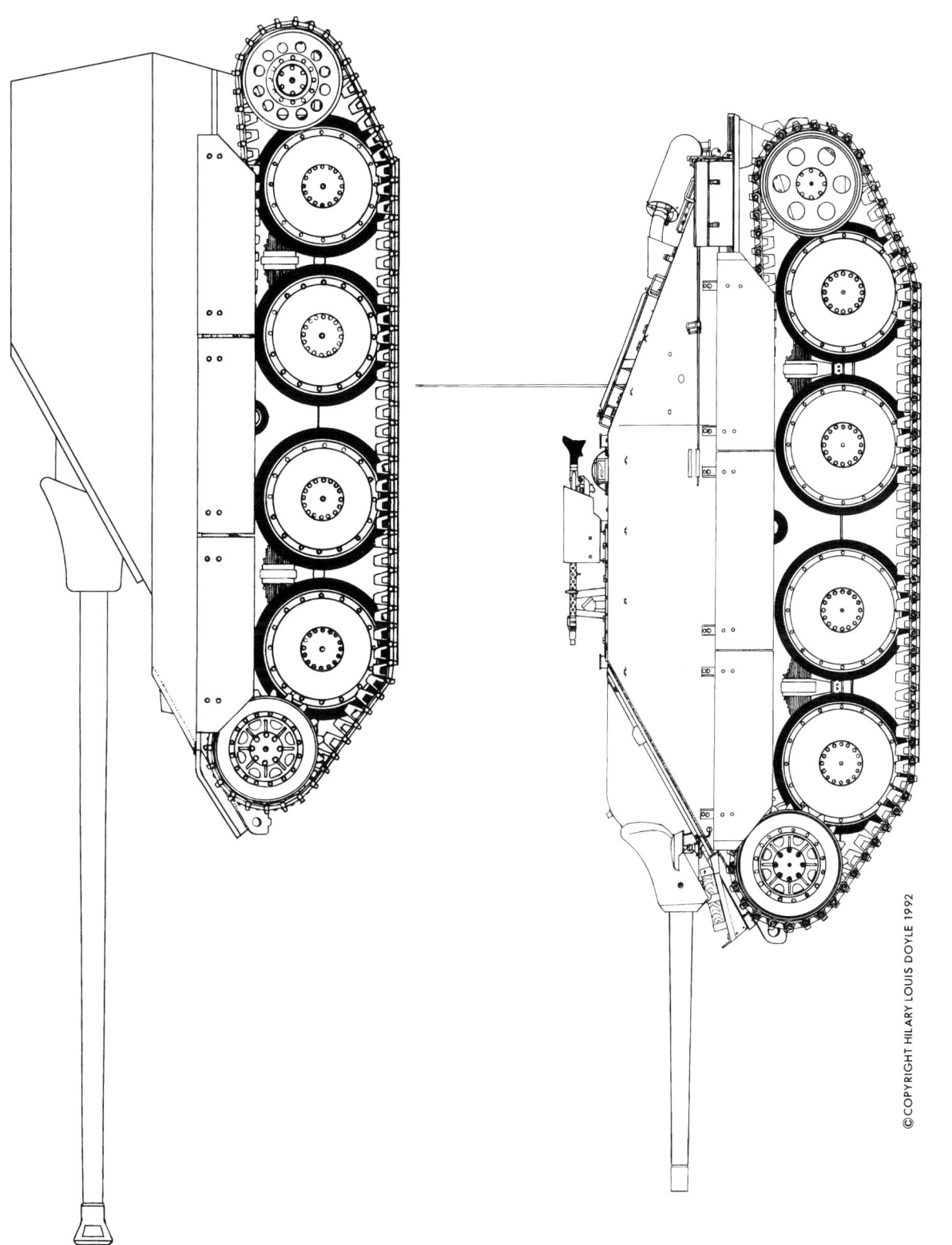

A comparison of a normal Jagdpanzer 38 (bottom) with Krupp's proposals in November 1944 to mount a Pz.Kpfw.IV turret on the chassis with a 7.5 cm Kw.K. L/48 or an 8 cm PAW 600 (page 50) or to create a Panzerjaeger 38t mit 7.5 cm L/70 (page 51)

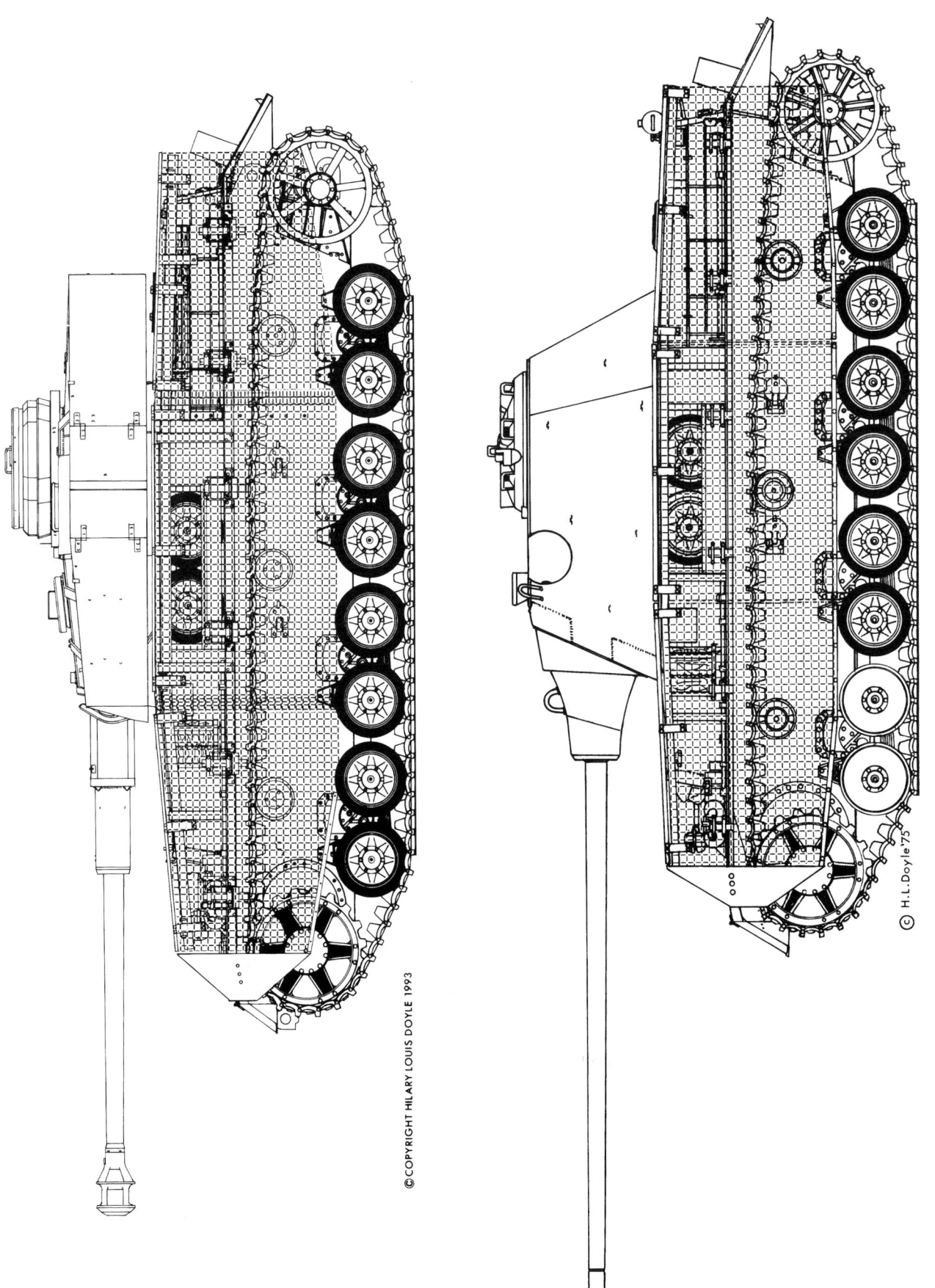

A comparison of a normal production Pz.Kpfw.IV Ausf.J with a 7.5 cm Kw.K. L/48 with Krupp's proposal in November 1944 to mount a Panther turret with a 7.5 cm Kw.K. L/70

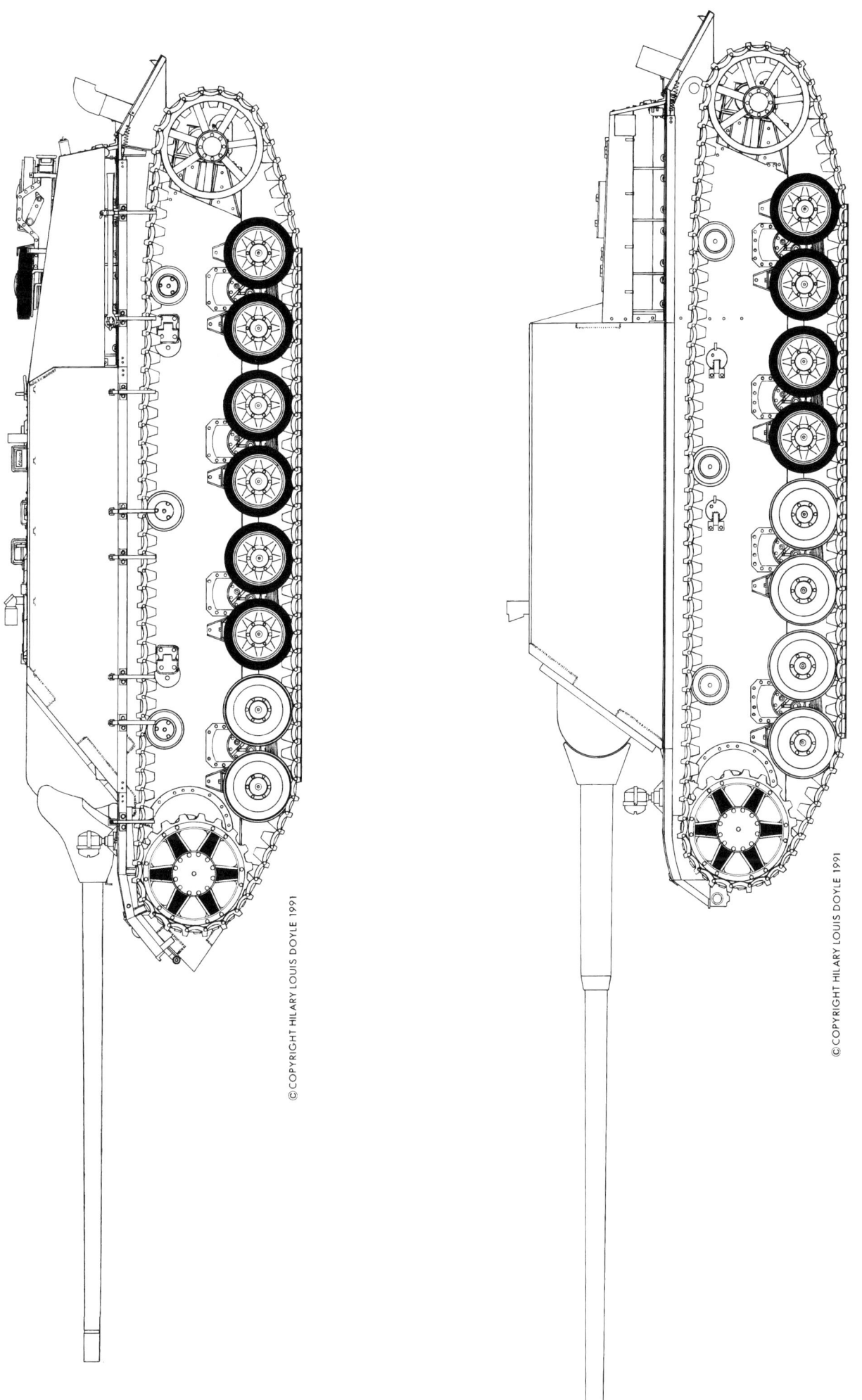

A comparison of a normal production Panzer IV/70 (V) with a 7.5 cm Pak L/70 with Krupp's proposal to mount an 8.8 cm Pak 43/2 L/71 in a superstructure on the same chassis

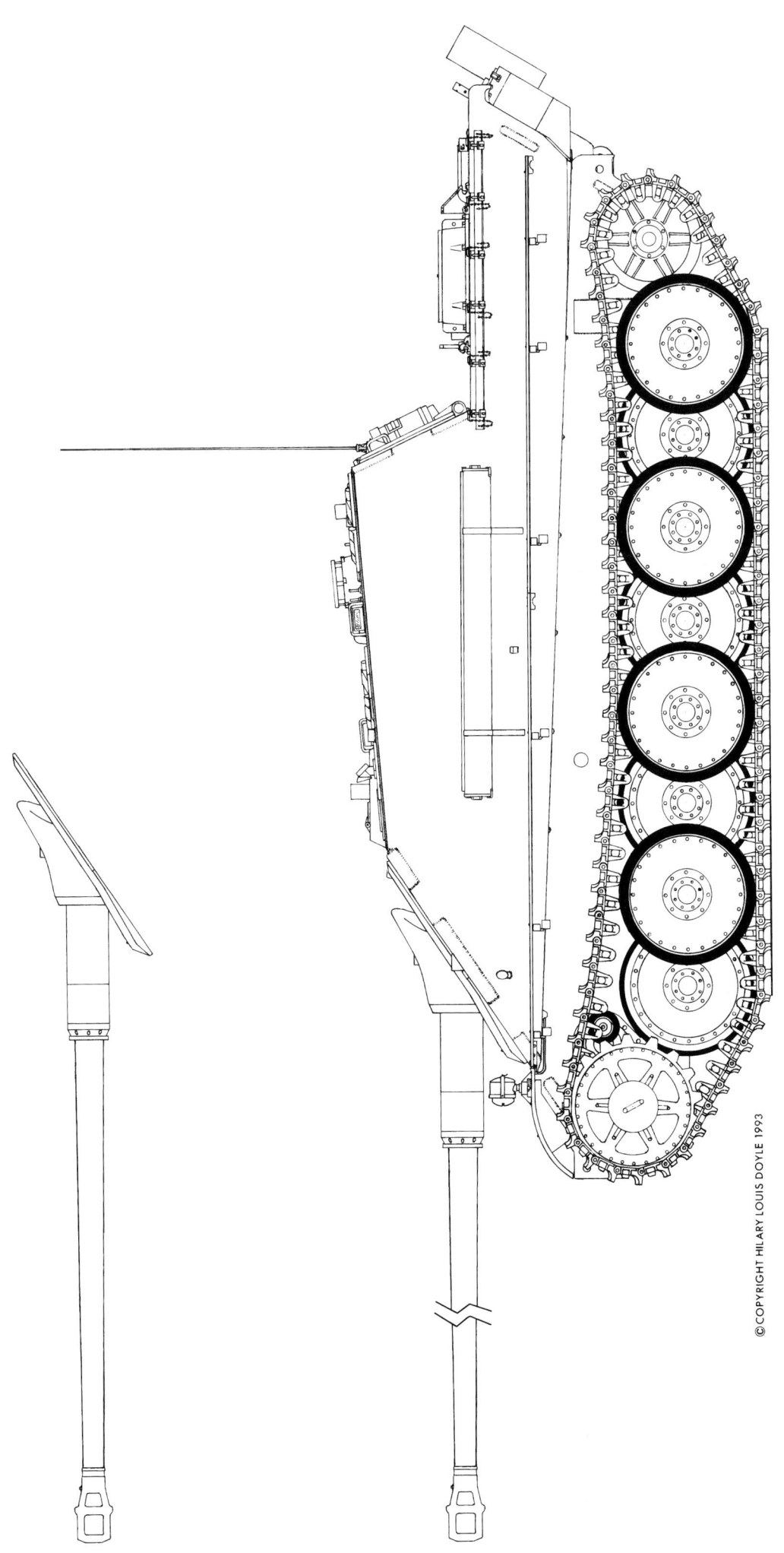

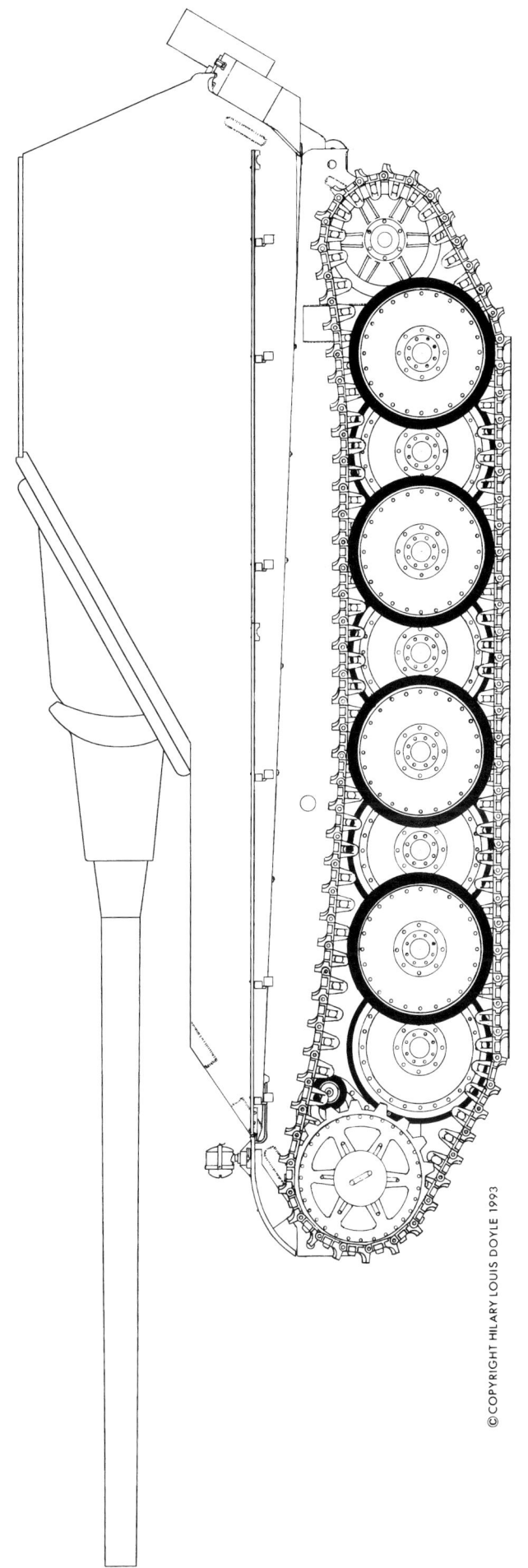

A comparison of a normal production Jagdpanther with an 8.8 cm Pak L/71 (page 54) with Krupp's proposal to mount a 12.8 cm Pak 80 L/55 at the rear (page 55)

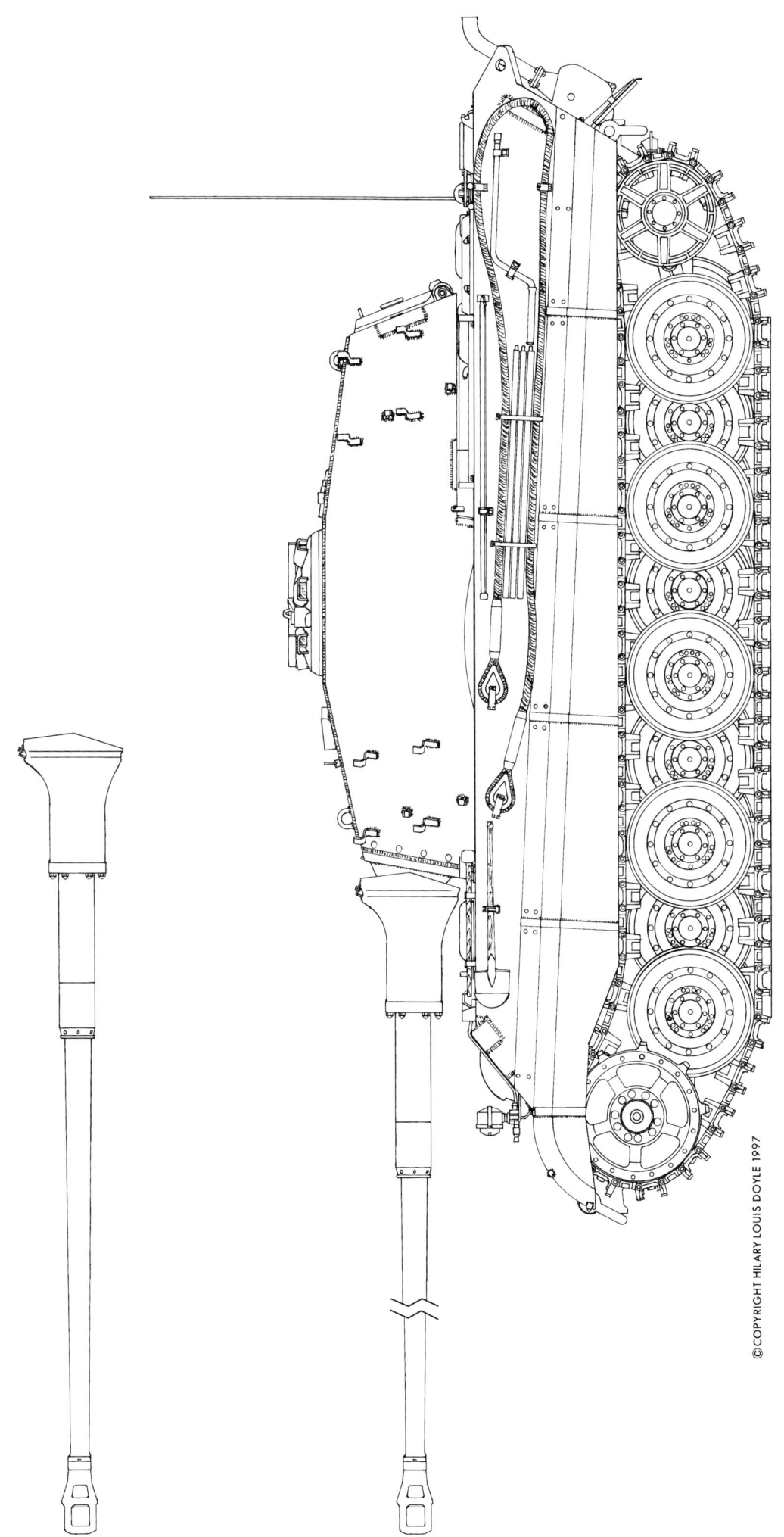

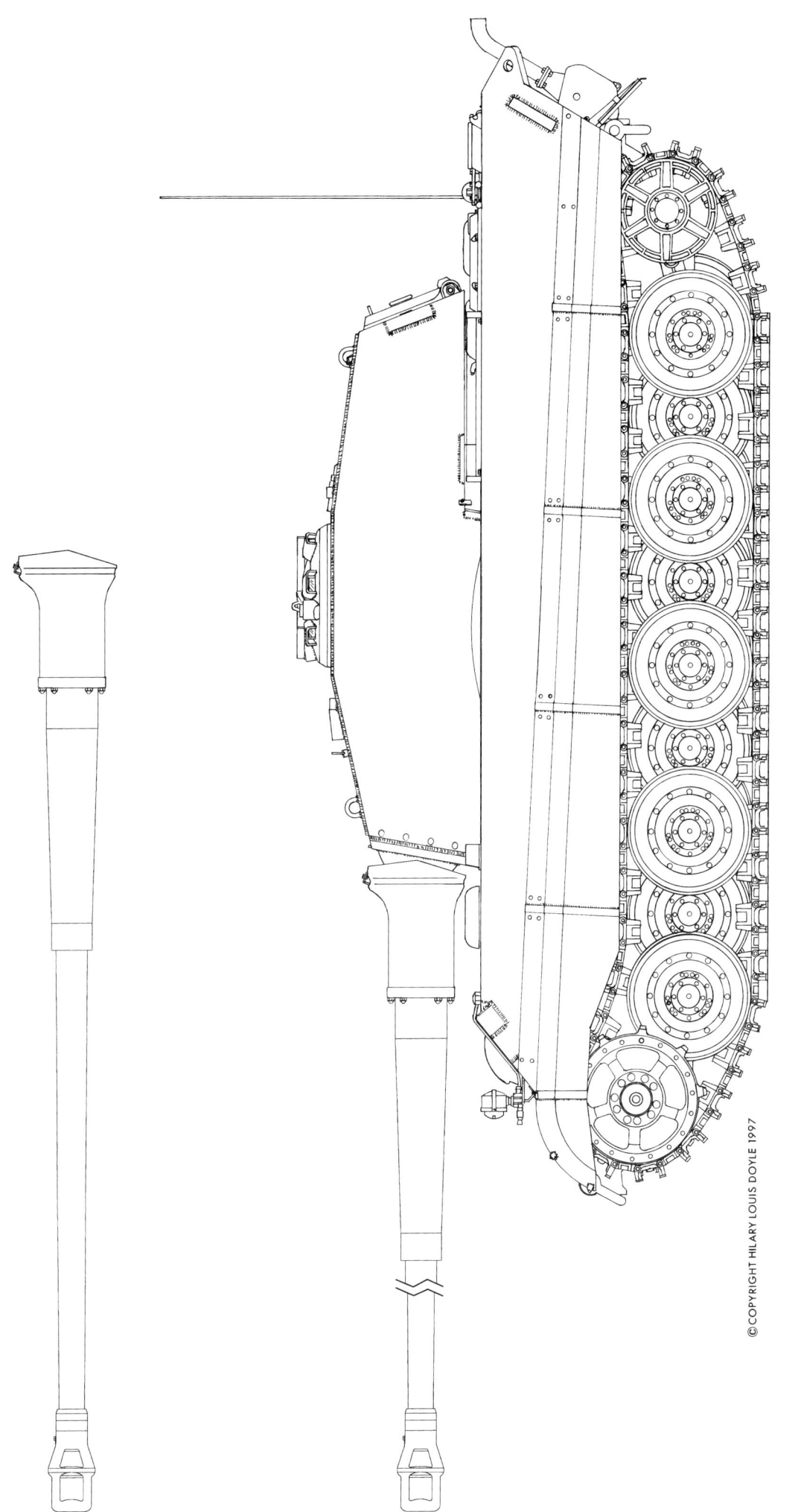

A comparison of a normal production Tiger II with an 8.8 cm Kw.K. L/71 (page 56) with Krupp's proposal in November 1944 to mount a 10.5 cm Kw.K. L/68 in the turret (page 57)

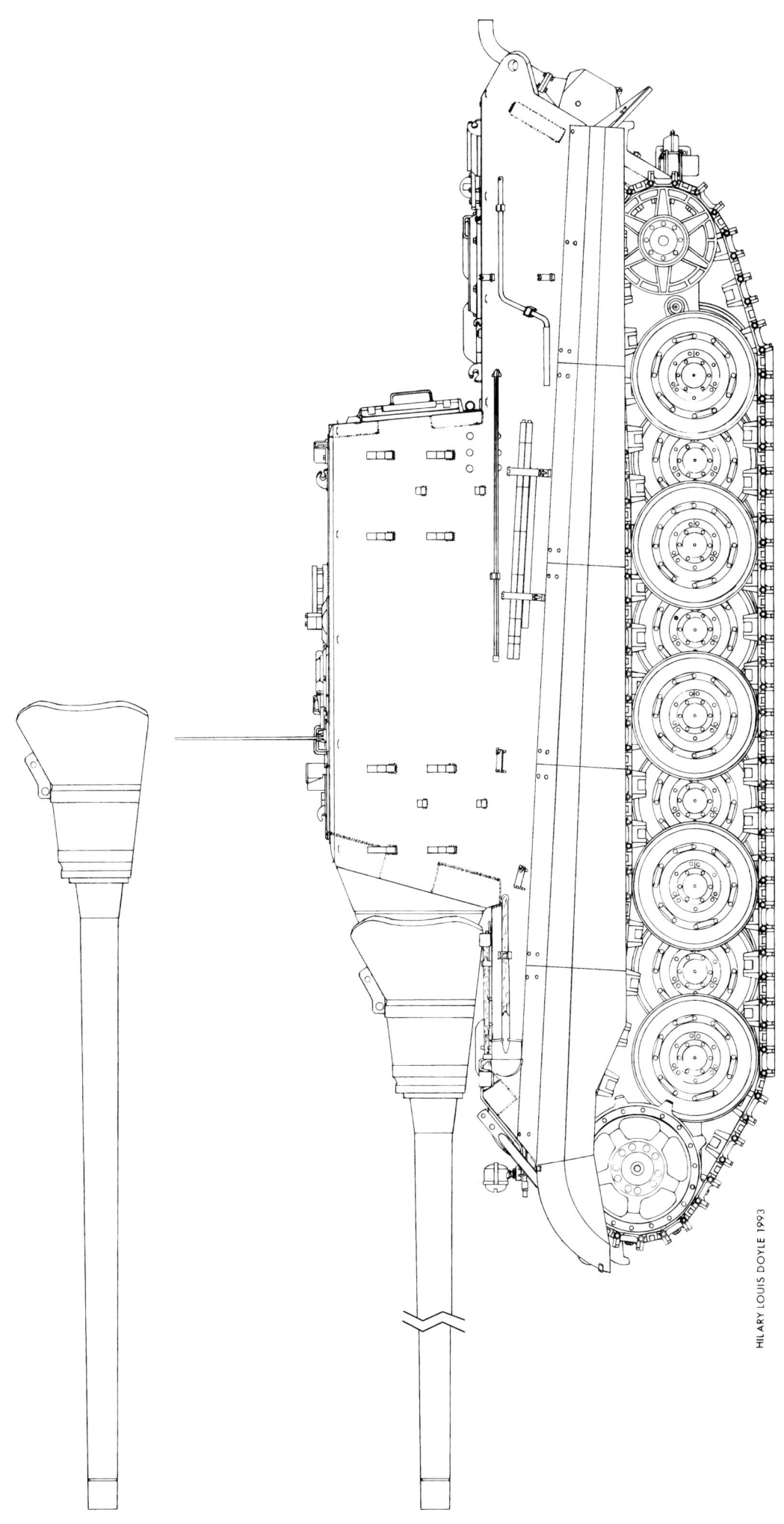

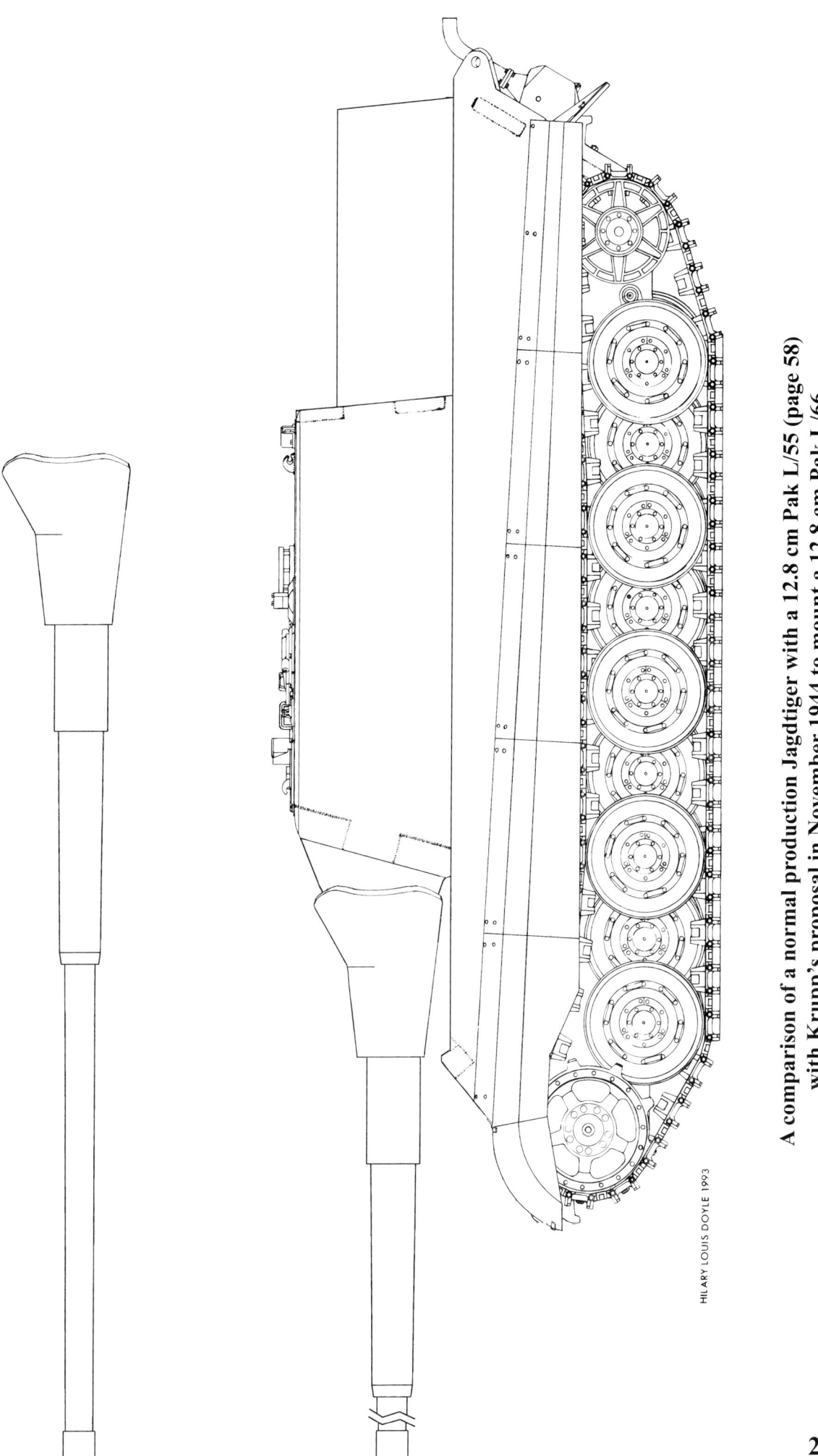

A comparison of a normal production Jagdtiger with a 12.8 cm Pak L/55 (page 58) with Krupp's proposal in November 1944 to mount a 12.8 cm Pak L/66 and to add an extension onto the superstructure rear to retract the gun (page 59)

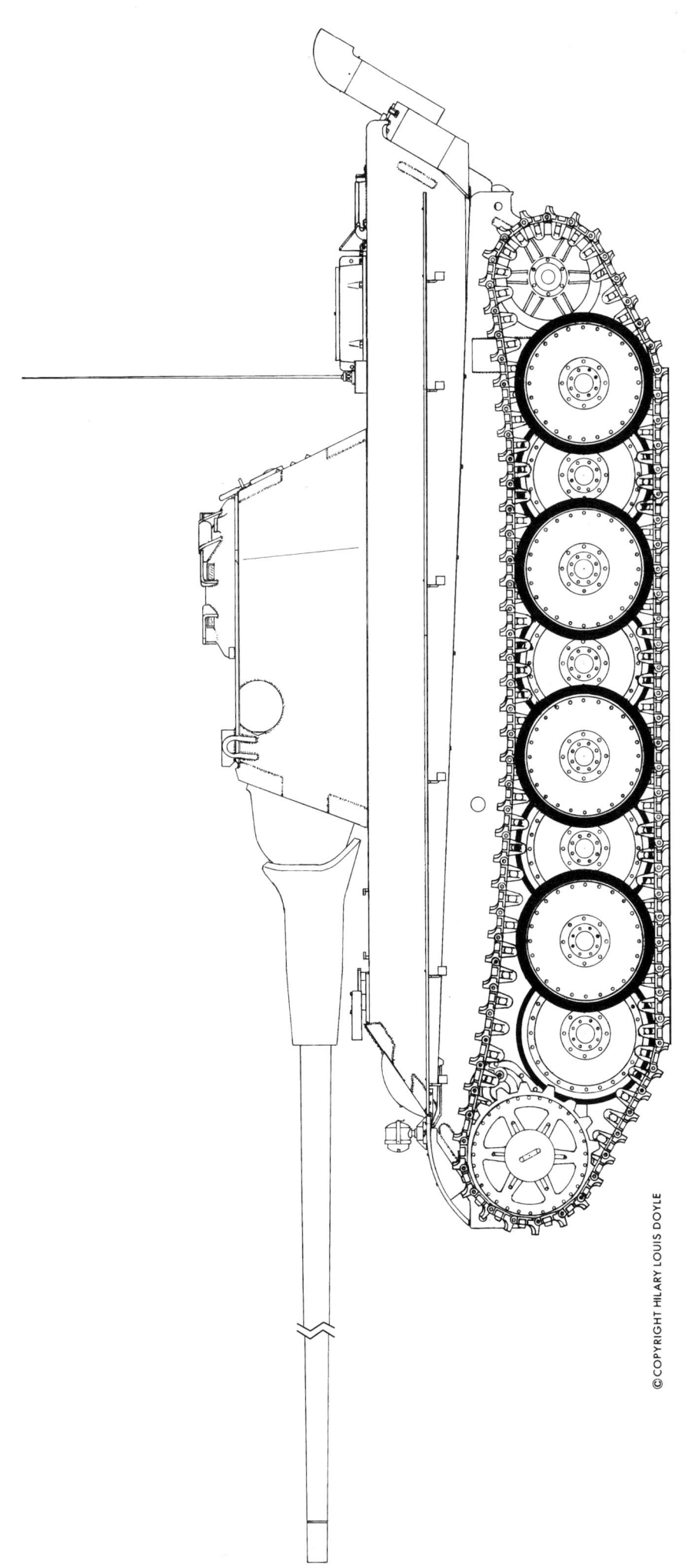

In November 1944, Krupp proposed to mount the 8.8 cm Kw.K. L/71 in a Panther Schmalturm (narrow turret) (refer to page 16 and 17)